Xiangjie Liu
Systems Control Theory

Also of Interest

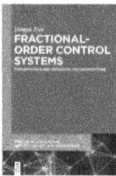

Fractional-Order Control Systems
D. Xue, 2017
ISBN 978-3-11-049999-5, e-ISBN (PDF) 978-3-11-049797-7,
e-ISBN (EPUB) 978-3-11-049719-9

Signals and Systems, Volumes 1+2
W. Zhang, 2017
ISBN 978-3-11-054409-1

Control Engineering
J. Sun, 2018
ISBN 978-3-11-057326-8, e-ISBN 978-3-11-057327-5,
e-ISBN (EPUB) 978-3-11-057336-7

Basic Process Engineering Control
P. Agachi, M. Cristea, 2014
ISBN 978-3-11-028981-7, e-ISBN (PDF) 978-3-11-028982-4,
e-ISBN (EPUB) 978-3-11-037701-9

Advanced Process Engineering Control
P. Agachi, M. Cristea, 2016
ISBN 978-3-11-030662-0, e-ISBN (PDF) 978-3-11-030663-7,
e-ISBN (EPUB) 978-3-11-038816-9

Xiangjie Liu

Systems Control Theory

—

DE GRUYTER

Science Press
Beijing

Author
Prof. Xiangjie Liu
China Science Publishing & Media Ltd.
16 Donghuangchenggen North Street
100717 Beijing
China
liuxj@ncepu.edu.cn

ISBN 978-3-11-057494-4
e-ISBN (PDF) 978-3-11-057495-1
e-ISBN (EPUB) 978-3-11-057527-9

Library of Congress Control Number: 2018014891

Bibliographic information published by the Deutsche Nationalbibliothek
The Deutsche Nationalbibliothek lists this publication in the Deutsche Nationalbibliografie;
detailed bibliographic data are available on the Internet at http://dnb.dnb.de.

© 2018 Walter de Gruyter GmbH, Berlin/Boston; Science Press
Typesetting: le-tex publishing services GmbH, Leipzig
Printing and binding: CPI books GmbH, Leck
Cover image: enot-poloskun/iStock/Getty Images

www.degruyter.com

Preface

"Modern Control Theory" is a fundamental and compulsory course for students who major in automation. The course takes place in the sixth semester. Together with the "automatic control theory" set up in the fifth term, it is the core theoretical basis of automation. During this period, the juniors already have an elementary knowledge of linear algebra, the Laplace transform and differential equations. The students also lay the solid foundation on feedback control theory. At this very opportune moment, students begin to learn modern control theory, which is more widely used in many aspects of modern control engineering.

This book is the result of teaching an undergraduate course over the years. The overall contents of the book can be described as follows. Chapter 1 introduces the system modeling with state space representation. Chapter 2 gives an overview of linear transformation of state vector. Chapter 3 presents solution of state space equations. Chapter 4 covers two types of stability for linear systems; namely, the I/O stability and the state related stability. Chapter 5 focuses on controllability and observability, with system decomposition and minimal realizations. Chapter 6 presents the state feedback and observer. The reader's understanding is developed further by experimenting with MATLAB command to develop simulations of their own control applications. Several benchmark problems on power generation modeling and control have also been incorporated into this book.

It is with gratitude that we acknowledge the continued support of the National Nature Science Foundation of China (61673171,61273144,60974051), Beijing Higher Education and Teaching Reformation projects (GJJG201409). We owe thanks to many colleagues and students. They frequently asked questions, pointed out problems, and, therefore, forced us to improve our work.

https://doi.org/10.1515/9783110574951-201

Contents

1 System Model

1.1 Introduction

In control theory research, the system model should be set up first. In this chapter, different kinds of models are discussed and some examples are given to show the reader how to set up a model of a system. The relationships between these model types are also described.

1.2 Models of Systems

A mathematical expression that appropriately relates the physical system quantities to the system components is called the mathematical model of a system.

There are, basically, two types of system descriptions; one is the external description, called the input-output description. The other is the internal one, called the state space description. In the former one, a system in operation involves the following three elements: the system's input (or excitation), the system itself, and the system's output (or response). This description just reveals the casual relationship between the external variables (the input and the output) without characterizing the internal structure. In the latter one, the description is a class of mathematic models based on the analysis of the internal structure of the system. It is a classical modern approach of describing a system. Correspondingly, there are two types of mathematical models of the system, which can facilitate the system analysis (it is well known that, in order to analyze a system, the mathematical model must be available).

The input-output description of the system will be introduced in Section 1.2.1 and 1.2.2, in the differential equation and the transfer function form. The state space model will be presented in Section 1.2.3.

1.2.1 Differential Equation

The differential equation is the fundamental mathematical model of a system. This description includes all the linearly independent equations of a system, as well as the appropriate initial conditions. The differential equation method is demonstrated by the following examples.

Example 1.1. Consider the network shown in Figure 1.1, where R, C, and L stand for the resistance, the capacitance and the inductance of the circuit respectively. Derive the network's differential equation mathematical model.

https://doi.org/10.1515/9783110574951-001

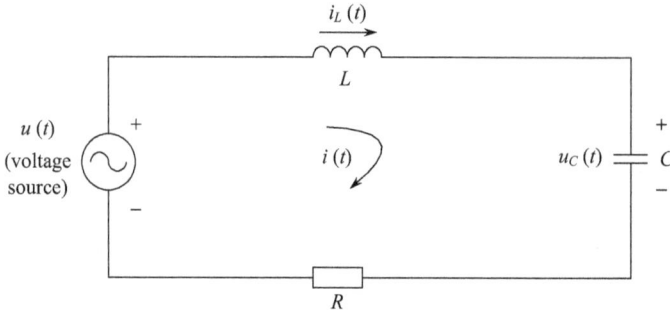

Fig. 1.1: RLC network.

Solution. Applying Kirchhoff's voltage law,

$$L\frac{di}{dt} + \frac{1}{C}\int_0^t idt + Ri = v(t) \tag{1.1}$$

The above integro-differential equation constitutes a mathematical description of the network. This model is a second order differential equation. Two appropriate initial conditions should be given to complete the description. The inductor's current $i_L(t)$ and the capacitor's voltage $vC(t)$ at the instant that the switch closes (at $t = 0$) are adopted as initial conditions:

$$i_L(0) = I_0$$
$$v_C(0) = V_0 ,$$

where I_0 and V_0 are given constants.

The integro-differential equation and the two initial conditions thus constitute a complete description of the network shown in Figure 1.1.

Example 1.2. Consider the network shown in Figure 1.2, where R, C and L stand for the resistance, the capacitance and the inductance of the circuit respectively. We assign the current of the inductance L_x ($x = 1, 2$) as i_x ($x = 1, 2$), and the voltage of the capacitance C_x ($x = 1, 2$) as v_x ($x = 1, 2$). Derive the network's differential equation mathematical model.

Solution. The differential equation method for describing this network is based on the three differential equations, which arise by applying Kirchhoff's current law. These three-loop equations are:

$$-L_1\frac{di_1}{dt} + u_{C1} + R_1 i_3 = 0 ,$$

$$-u_{C1} + L_2\frac{di_2}{dt} + R_2 i_4 = 0 , \tag{1.2a}$$

$$L_2\frac{di_2}{dt} - L_1\frac{di_1}{dt} - u_{C2} = 0 .$$

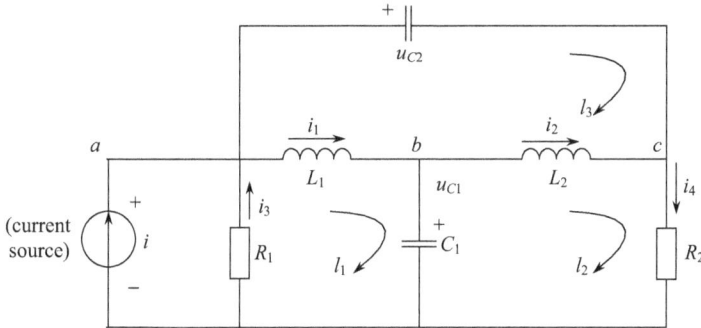

Fig. 1.2: A three-loop network.

By applying Kirchhoff's voltage law, we have:

$$i + i_3 + i_1 - C_2 \frac{du_{C2}}{dt} = 0 \,,$$

$$C_1 \frac{du_{C1}}{dt} + i_1 + i_2 = 0 \,, \tag{1.2b}$$

$$C_2 \frac{di_2}{dt} + i_2 - i_4 = 0 \,.$$

The initial conditions are $v_{C_1}(0) = v_{C10}$, $v_{C_2}(0) = v_{C20}$ and $i_{L_1}(0) = i_{L_10}$, $i_{L_1}(0) = i_{L_10}$.

Example 1.3. Consider the mechanical system shown in Figure 1.3, where y, K, m and B are the position of the mass, the spring's constant, the mass, and the friction coefficient respectively. Derive the system's differential equation mathematical model.

Solution. By using d'Alembert's law of forces, the following differential equation is obtained:

$$m \frac{d^2y}{dt^2} + B \frac{dy}{dt} + Ky = f(t) \,. \tag{1.3}$$

The initial conditions of the above equation are the distance $y(t)$ and the velocity $v(t) = dy/dt$ at the instant $t = 0$, i.e., at the instant when the external force $f(t)$ is applied. Therefore, the initial conditions are:

$$y(0) = Y_0 \quad \text{and} \quad v(0) = \left[\frac{dy}{dt} \right]_{t=0} = V_0 \,,$$

where Y_0 and V_0 are given constants.

The differential equations and the two initial conditions constitute the complete description of the mechanical system shown in Figure 1.3.

Remark 1.2.1. A differential equation is a description in the time domain, which can be applied to many categories of systems, such as linear and nonlinear systems, time invariant and time variant systems with lumped and distributed parameters, and zero and nonzero initial conditions, among many others.

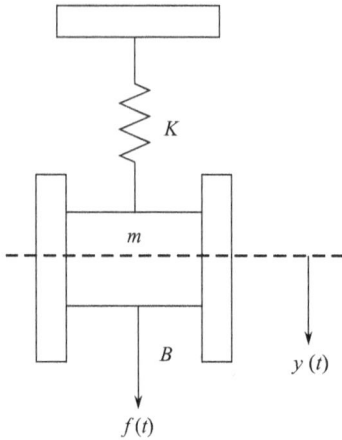

Fig. 1.3: A spring and a mass.

1.2.2 Transfer Function

In contrast to the differential equation, which is a description in the time domain, the transfer function model is a description in the Laplace domain and holds only for a restricted category of systems, i.e., for linear time invariant (LTI) systems with zero initial conditions. The transfer function is designated by $G(s)$ and is defined as follows.

Definition. The transfer function $G(s)$ of a linear, time invariant system with zero initial conditions is the ratio of the Laplace transform of the output $y(t)$ to the Laplace transform of the input $u(t)$, i.e.,

$$G(s) = \frac{L\{y(t)\}}{L\{u(t)\}} = \frac{Y(s)}{U(s)} \,. \tag{1.4}$$

The introductory examples used in Section 1.2.1 will also be used for the derivation of their transfer functions.

Example 1.4. Consider the network shown in Figure 1.1. Derive the transfer function $G(s) = I(s)/V(s)$.

Solution. Figure 1.1, in the Laplace domain and with zero initial conditions I_0 and V_0, can be shown in Figure 1.4. From Kirchhoff's voltage law,

$$LsI(s) + RI(s) + \frac{I(s)}{Cs} = V(s) \,. \tag{1.5}$$

The transfer function is:

$$G(s) = \frac{I(s)}{V(s)} = \frac{I(s)}{\left[Ls + R + \frac{1}{Cs}\right] I(s)} = \frac{Cs}{LCs^2 + RCs + 1} \,. \tag{1.6}$$

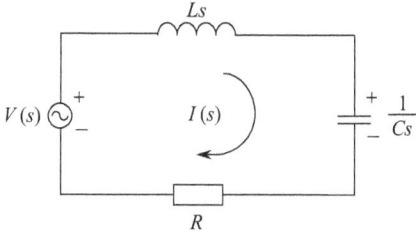

Fig. 1.4: RLC circuit.

When the voltage $V_R(s)$ across the resistor is chosen as the output, the transfer function becomes:

$$G(s) = \frac{V_R(s)}{V(s)} = \frac{RI(s)}{V(s)} = \frac{RCs}{LCs^2 + RCs + 1} \, . \tag{1.7}$$

Example 1.5. Consider the electrical network shown in Figure 1.2. Determine the transfer function $G(s) = I_2(s)/V(s)$.

Solution. This network in the Laplace domain, with zero initial conditions, is shown in Figure 1.5. The equations for the two loops can be expressed as:

$$\left[R_1 + \frac{1}{Cs}\right] I_1(s) - \frac{1}{Cs} I_2(s) = V(s) \tag{1.8a}$$

$$-\frac{1}{Cs} I_1(s) + \left[R_2 + Ls + \frac{1}{Cs}\right] I_2(s) = 0 \, . \tag{1.8b}$$

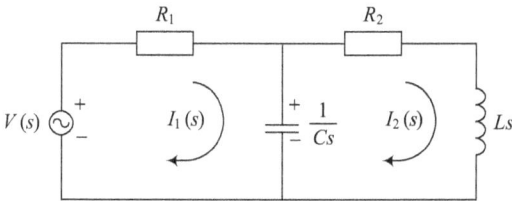

Fig. 1.5: A two-loop network.

Equation (1.8b) yields:

$$I_1(s) = \left[LCs^2 + R_2 Cs + 1\right] I_2(s) \, . \tag{1.9}$$

Substituting equation (1.9) into equation (1.8a), we get:

$$(R_1 Cs + 1)(LCs^2 + R_2 Cs + 1)I_2(s) - I_2(s) = CsV(s) \, . \tag{1.10}$$

Hence,

$$G(s) = \frac{I_2(s)}{V(s)} = \frac{Cs}{(R_1 Cs + 1)(LCs^2 + R_2 Cs + 1) - 1}$$

$$= \frac{1}{R_1 LCs + (R_1 R_2 C + L)s + R_1 + R_2} \, . \tag{1.11}$$

Example 1.6. Consider the mechanical system shown in Figure 1.3. Determine the transfer function $G(s) = Y(s)/F(s)$.

Solution. This system, in the Laplace domain and with zero initial conditions, is shown in Figure 1.6.

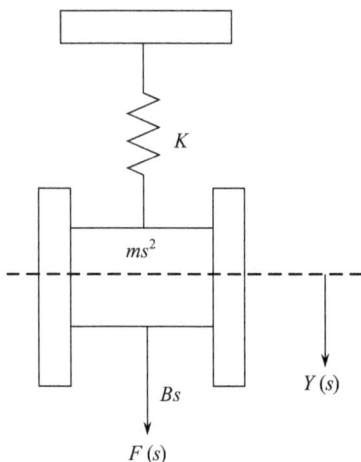

Fig. 1.6: A spring and a mass.

Using d'Alembert's law of forces, we get:

$$ms^2 Y(s) + Bs Y(s) + KY(s) = F(s) . \tag{1.12}$$

The transfer function is:

$$G(s) = \frac{Y(s)}{F(s)} = \frac{1}{ms^2 + Bs + K} . \tag{1.13}$$

Remark 1.2.2. In the above examples, it can be seen that the transfer function $G(s)$ is the ratio of two polynomials in the Laplace domain. In general, $G(s)$ has the following form:

$$G(s) = \frac{\beta_m s^m + \beta_{m-1} s^{m-1} + \cdots + \beta_1 s + \beta_0}{s^n + \alpha_{n-1} s^{n-1} + \cdots + \alpha_1 s + \alpha_0} = K \frac{\prod_{i=1}^{m}(s + z_i)}{\prod_{i=1}^{n}(s + p_i)} , \tag{1.14}$$

where $-p_i$ ($i = 1, 2, \ldots, n$) are the roots of the denominator, which are called the poles of $G(s)$, and $-z_i$ are the roots of the numerator, which are called the zeros of $G(s)$. Poles and zeros (particularly the poles) play a significant role in the behavior of a system.

1.2.3 The State Space Model

The state space model is a description in the time domain, which may be applied to a very wide category of systems, such as linear and nonlinear systems, time invariant and time variant systems, systems with nonzero initial conditions, etc. The *state* of a system refers to the past, present, and future of the system. From a mathematical point of view, the *state* of a system is expressed by its state variables. Usually, a system is described by a finite number of state variables, which are designated by $x_1(t), x_2(t), \ldots, x_n(t)$ and are defined as follows.

1.2.3.1 Definition

The state variables $x_1(t), x_2(t), \ldots, x_n(t)$ of a system are defined as a (minimum) number of variables, such that, if we know the following, the determination of the system's states for $t > t_0$ is guaranteed:
(1) their values at a certain instant t_0
(2) the input of the system for $t \geq t_0$
(3) the mathematical model, which relates the inputs, the state variables, and the system itself

Consider a system with multiple inputs and multiple outputs (MIMO), as shown in Figure 1.7.

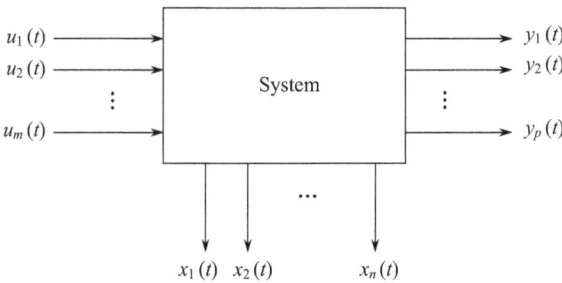

Fig. 1.7: System with multiple inputs and multiple outputs (MIMO).

The input vector is designated by $u(t)$ and has the form:

$$u(t) = \begin{bmatrix} u_1(t) \\ u_2(t) \\ \vdots \\ u_m(t) \end{bmatrix}, \qquad (1.15)$$

where m is the number of inputs. The output vector is designated by $y(t)$ and has the form:

$$y(t) = \begin{bmatrix} y_1(t) \\ y_2(t) \\ \vdots \\ y_p(t) \end{bmatrix}, \qquad (1.16)$$

where p is the number of outputs. The *state vector* $x(t)$ has the form:

$$x(t) = \begin{bmatrix} x_1(t) \\ x_2(t) \\ \vdots \\ x_n(t) \end{bmatrix}, \tag{1.17}$$

where n is the number of state variables.

The *state equations* are a number of n first order differential equations, which relate the input vector $u(t)$ to the state vector $x(t)$ and have the form:

$$\dot{x}(t) = f[x(t), u(t)] , \tag{1.18}$$

where $f(\cdot)$ is a column with n elements. The function $f(\cdot)$, in general, is a complex nonlinear function of $x(t)$ and $u(t)$. Note that equation (1.18) is a set of *dynamic equations*.

The output vector $y(t)$ of the system is related to the input vector $u(t)$ and the state vector $x(t)$ as follows:

$$y(t) = g[x(t), u(t)] , \tag{1.19}$$

where $g(\cdot)$ is a column with p elements. Relation (1.19) is called the *output equation*. The function $g(\cdot)$ is generally a complex nonlinear function of $x(t)$ and $u(t)$. Note that equation (1.19) is a set of *algebraic (nondynamic) equations*.

The initial conditions of the state space equation (1.18) are the values of the elements of the state vector $x(t)$ for $t = t_0$, and is denoted as:

$$x(t_0) = x_0 = \begin{bmatrix} x_1(t_0) \\ x_2(t_0) \\ \vdots \\ x_n(t_0) \end{bmatrix}. \tag{1.20}$$

The state space equation (1.18), the output equation (1.19), and the initial conditions (1.20), i.e., the following equations, constitute the description of a dynamic system in the *state space*.

$$\dot{x}(t) = f[x(t), u(t)] , \tag{1.21a}$$
$$y(t) = g[x(t), u(t)] , \tag{1.21b}$$
$$x(t_0) = x_0 . \tag{1.21c}$$

Since the dynamic state equation (1.21a) plays a dominant role in equations (1.21), all the three equations in (1.21), will be called, for simplicity, state equations.

When the system is a linear stationary one, the state space model is:

$$\dot{x}(t) = Ax(t) + Bu(t)$$
$$y(t) = Cx(t) + Du(t),$$

(1.22)

where A is the systematic matrix; B is the input/control matrix; C is the output matrix, and D is the direct transfer matrix.

The state equations (1.21) are, in the field of automatic control, the modern method of system description. Thus, the state space model relates the following four elements: the input, the system, the state variables, and the output. In contrast, the differential equations and the transfer function relate three elements: the input, the system and the output – wherein the input is related to the output via the system directly (i.e., without giving information about the state of the system). It is exactly for this reason that these two models are called input-output models.

2. The Construction of the State Space Model
There are three ways to set up the state space model, which can be based on:
(1) the transform of the block diagram
(2) first principal modeling
(3) the input-output model

Each way will be described in detail and examples will be given.

The Transform of the Block Diagram
Example 1.7. The system block diagram is shown in Figure 1.8 (a). u is the input and y is the output. Try to deduce the state space equations.

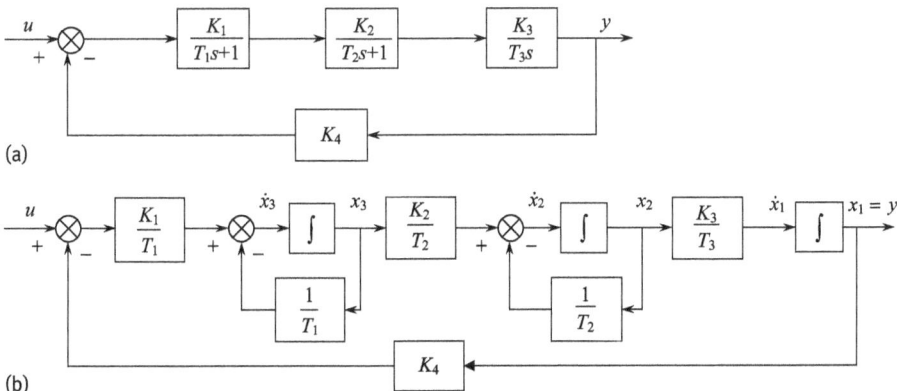

Fig. 1.8: System block diagram.

Solution. The structure of each link part is shown in Figure 1.8 (b), thus the relative equations can be derived as follows.

The state equations:

$$\begin{cases} \dot{x}_1 = \frac{K_3}{T_3}x_2 \\ \dot{x}_2 = -\frac{1}{T_2}x_2 + \frac{K_2}{T_2}x_3 \\ \dot{x}_3 = -\frac{1}{T_1}x_3 - \frac{K_1 K_4}{T_1}x_1 + \frac{K_1}{T_1}u \ . \end{cases} \tag{1.23}$$

The output equation:

$$y = x_1 \ .$$

The above equations can be rewritten in the vector matrix form:

$$\dot{x} = \begin{bmatrix} 0 & \frac{K_3}{T_3} & 0 \\ 0 & -\frac{1}{T_2} & \frac{K_2}{T_2} \\ -\frac{K_1 K_4}{T_1} & 0 & -\frac{1}{T_1} \end{bmatrix} x + \begin{bmatrix} 0 \\ 0 \\ \frac{K_1}{T_1} \end{bmatrix} u \tag{1.24}$$

$$y = \begin{bmatrix} 1 & 0 & 0 \end{bmatrix} x \ .$$

The First Principal Modeling
When a physical system is given, the mechanism analysis can be carried out with proper assumption and simplification. The mechanism model can be set up with chosen inputs and outputs. If the middle variables are eliminated, then the differential equation mentioned above can be obtained. If the middle variables are chosen as the state variables, then the state space model can be achieved.

Example 1.8. Consider the system shown in Figure 1.9. The current of $C_{1,2}$ is $C_{1,2}\dot{u}_{C1,2}$ respectively, and the voltage of $L_{1,2}$ is $L_{1,2}\dot{i}_{1,2}$. The input is the current source, and the outputs are the voltages of capacitances, C_1 and C_2. Derive the system's state space representation.

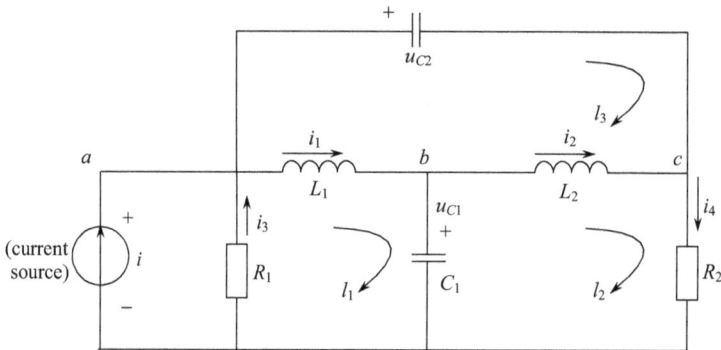

Fig. 1.9: A three-loop network.

Solution. The differential equations method for describing this network is based on the three differential equations, which arise by applying Kirchhoff's voltage law. These three-loop equations are:

$$-L_1\frac{di_1}{dt} + u_{C1} + R_1 i_3 = 0,$$

$$-u_{C1} + L_2\frac{di_2}{dt} + R_2 i_4 = 0,$$

$$L_2\frac{di_2}{dt} - L_1\frac{di_1}{dt} - u_{C2} = 0.$$

By applying Kirchhoff's current law, we get:

$$i + i_3 + i_1 - C_2\frac{du_{C2}}{dt} = 0,$$

$$C_1\frac{du_{C1}}{dt} + i_1 + i_2 = 0,$$

$$C_2\frac{du_{C2}}{dt} + i_2 - i_4 = 0.$$

Define:

$$u_{C_1} = x_1, \qquad u_{C_2} = x_2,$$

$$i_1 = x_3, \qquad i_2 = x_4.$$

The system's state space equation can be expressed as:

$$
\begin{pmatrix} \dot{x}_1 \\ \dot{x}_2 \\ \dot{x}_3 \\ \dot{x}_4 \end{pmatrix}
=
\begin{pmatrix}
0 & 0 & -\frac{1}{C_1} & -\frac{1}{C_1} \\
0 & -\frac{1}{C_2(R_1+R_2)} & \frac{R_1}{C_2(R_1+R_2)} & -\frac{R_2}{C_2(R_1+R_2)} \\
\frac{1}{L_1} & -\frac{R_1}{L_1(R_1+R_2)} & -\frac{R_1 R_2}{L_1(R_1+R_2)} & -\frac{R_1 R_2}{L_1(R_1+R_2)} \\
\frac{1}{L_2} & -\frac{R_2}{L_2(R_1+R_2)} & -\frac{R_1 R_2}{L_2(R_1+R_2)} & -\frac{R_1 R_2}{L_2(R_1+R_2)}
\end{pmatrix}
\begin{pmatrix} x_1 \\ x_2 \\ x_3 \\ x_4 \end{pmatrix}
+
\begin{pmatrix}
0 \\
\frac{R_1}{C_2(R_1+R_2)} \\
-\frac{R_1 R_2}{L_1(R_1+R_2)} \\
-\frac{R_1 R_2}{L_2(R_1+R_2)}
\end{pmatrix} i,
$$

$$(1.25)$$

$$
\begin{pmatrix} y_1 \\ y_2 \end{pmatrix}
=
\begin{pmatrix} u_{C_1} \\ u_{C_2} \end{pmatrix}
=
\begin{pmatrix} 1 & 0 & 0 & 0 \\ 0 & 1 & 0 & 0 \end{pmatrix}
\begin{pmatrix} x_1 \\ x_2 \\ x_3 \\ x_4 \end{pmatrix}.
$$

Example 1.9. Consider the mechanical system shown in Figure 1.10, where y, K, m and B are the position of the mass, the spring's constant, the mass, and the friction coefficient respectively. In the role of external forces f, derive the system's state space representation where y_1, y_2 are the outputs.

Solution. Choose the position y_1, y_2 and the velocity v_1, v_2 of the mass M_1, M_2 as the state variables:

$$x_1 = y_1, \qquad\qquad x_2 = y_2,$$

$$x_3 = v_1 = \frac{dy_1}{dt}, \qquad x_4 = v_2 = \frac{dy_2}{dt}.$$

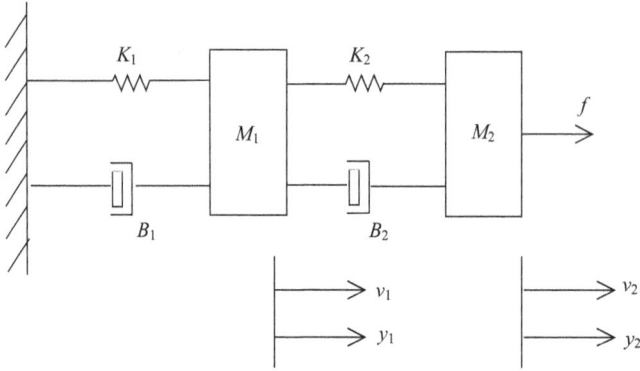

Fig. 1.10: The mass-spring-damper system.

By using Newton's laws of motion, for M_1 we get:

$$M_1 \frac{dv_1}{dt} = K_2(y_2 - y_1) + B_2 \left(\frac{dy_2}{dt} - \frac{dy_1}{dt} \right) - K_1 y_1 - B_1 \frac{dy_1}{dt} \,.$$

For M_2:

$$M_2 \frac{dv_2}{dt} = f - K_2(y_2 - y_1) - B_2 \left(\frac{dy_2}{dt} - \frac{dy_1}{dt} \right) \,.$$

With $u = f$, we can get:

$$\dot{x}_1 = x_3$$

$$\dot{x}_2 = x_4$$

$$\dot{x}_3 = -\frac{1}{M_1}(K_1 + K_2)x_1 + \frac{K_2}{M_1}x_2 - \frac{1}{M_1}(B_1 + B_2)x_3 + \frac{B_2}{M_1}x_4$$

$$\dot{x}_4 = \frac{K_2}{M_2}x_1 - \frac{K_2}{M_2}x_2 + \frac{B_2}{M_2}x_3 - \frac{B_2}{M_2}x_4 \,.$$

The above equations can be expressed in a compact form:

$$\begin{pmatrix} \dot{x}_1 \\ \dot{x}_2 \\ \dot{x}_3 \\ \dot{x}_4 \end{pmatrix} = \begin{pmatrix} 0 & 0 & 1 & 0 \\ 0 & 0 & 0 & 1 \\ -\frac{1}{M_1}(K_1 + K_2) & \frac{K_2}{M_1} & -\frac{1}{M_1}(B_1 + B_2) & \frac{B_2}{M_1} \\ \frac{K_2}{M_2} & -\frac{K_2}{M_2} & \frac{B_2}{M_2} & -\frac{B_2}{M_2} \end{pmatrix} \begin{pmatrix} x_1 \\ x_2 \\ x_3 \\ x_4 \end{pmatrix} + \begin{pmatrix} 0 \\ 0 \\ 0 \\ \frac{1}{M_2} \end{pmatrix} f \,.$$

$$(1.26)$$

The output equation is:

$$\begin{pmatrix} y_1 \\ y_2 \end{pmatrix} = \begin{pmatrix} 1 & 0 & 0 & 0 \\ 0 & 1 & 0 & 0 \end{pmatrix} \begin{pmatrix} x_1 \\ x_2 \\ x_3 \\ x_4 \end{pmatrix} \,.$$

Example 1.10. Consider a cart with an inverted pendulum hinged on top of it, as shown in Figure 1.11. For simplicity, the cart and the pendulum are assumed to move in only one plane, while the friction, the mass of the stick, and the gust of wind are disregarded. The problem is to maintain the pendulum at the vertical position.

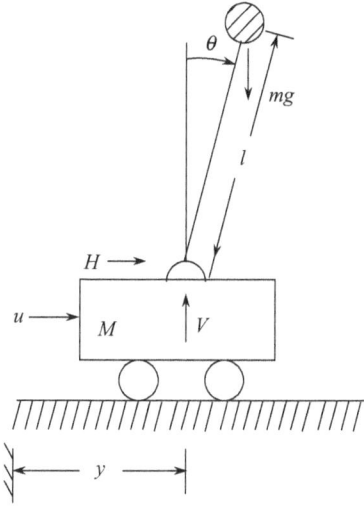

Fig. 1.11: The inverted pendulum system.

Suppose H and V are, respectively, the horizontal and vertical forces exerted by the cart on the pendulum as shown. The application of Newton's law to the linear movements yields:

$$M\frac{d^2y}{dt^2} = u - H ,$$

$$H = m\frac{d^2}{dt^2}(y + l\sin\theta) = m\ddot{y} + ml\ddot{\theta}\cos\theta - ml(\dot{\theta})^2\sin\theta ,$$

$$mg - V = m\frac{d^2}{dt^2}(l\cos\theta) = ml[-\ddot{\theta}\sin\theta - (\dot{\theta})^2\cos\theta] .$$

The application of Newton's law to the rotational movement of the pendulum around the hinge yields:

$$mgl\sin\theta = ml\ddot{\theta}\cdot l + m\ddot{y}l\cos\theta ,$$

$$\sin\theta = \theta , \quad \cos\theta = 1 ,$$

$$mg = V ,$$

$$M\ddot{y} = u - m\ddot{y} - ml\ddot{\theta} , \quad g\theta = l\ddot{\theta} + \ddot{y} ,$$

which imply:

$$M\ddot{y} = u - mg\theta ,$$

$$Ml\ddot{\theta} = (M + m)g\theta - u .$$

Define:

$$x_1 = y, \quad x_2 = \dot{y}, \quad x_3 = \theta, \quad x_4 = \dot{\theta}.$$

Then the state space model can be derived as:

$$\begin{bmatrix} \dot{x}_1 \\ \dot{x}_2 \\ \dot{x}_3 \\ \dot{x}_4 \end{bmatrix} = \begin{bmatrix} 0 & 1 & 0 & 0 \\ 0 & 0 & \frac{-mg}{M} & 0 \\ 0 & 0 & 0 & 1 \\ 0 & 0 & \frac{(M+m)g}{Ml} & 0 \end{bmatrix} \begin{bmatrix} x_1 \\ x_2 \\ x_3 \\ x_4 \end{bmatrix} + \begin{bmatrix} 0 \\ \frac{1}{M} \\ 0 \\ \frac{-1}{Ml} \end{bmatrix} u, \tag{1.27}$$

$$y = \begin{bmatrix} 1 & 0 & 0 & 0 \end{bmatrix} x.$$

Example 1.11. Consider a separately excited DC motor system, as shown in Figure 1.12. In the diagram, R and L stand for the resistance and inductance of the armature loop respectively. J is the inertia of the rotating part, and B is the viscous friction coefficient. Develop the state space equations when the armature voltage u is chosen as the control variable.

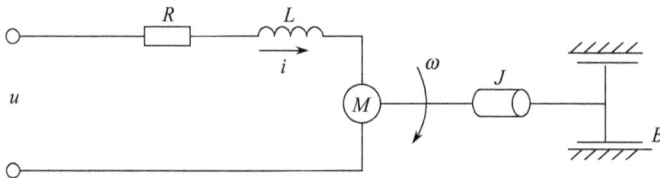

Fig. 1.12: The system of a separately excited DC motor.

Since the inductance L and the rotating inertia J are energy storage elements, their corresponding physical variables, e.g., the current i and the rotating angular speed ω, are independent of each other. They could be chosen as state variables:

$$x_1 = i,$$
$$x_2 = \omega,$$

then as:

$$\frac{dx_1}{dt} = \frac{di}{dt}, \quad \frac{dx_2}{dt} = \frac{d\omega}{dt}.$$

Using the circuit equations of the armature circuit,

$$L\frac{di}{dt} + Ri + e = u.$$

According to the dynamics equations, we have:

$$J\frac{d\omega}{dt} + B\omega = K_a i.$$

According to the electromagnetic induction relationship, we get:

$$e = K_b \omega \,,$$

where e is the back electromotive force and K_a, K_b are the torque constant and back electromotive force respectively.

According to the three equations above, the model of the system may be rewritten as:

$$\frac{di}{dt} = -\frac{R}{L} i + -\frac{K_b}{L} \omega + \frac{1}{L} u \,,$$

$$\frac{d\omega}{dt} = \frac{K_a}{J} i - \frac{B}{J} \omega \,.$$

By putting $x_1 = i$, $x_2 = \omega$ into the above equations, we get:

$$\begin{pmatrix} \dot{x}_1 \\ \dot{x}_2 \end{pmatrix} = \begin{pmatrix} -\frac{R}{L} & -\frac{K_b}{L} \\ \frac{K_a}{J} & -\frac{B}{J} \end{pmatrix} \begin{pmatrix} x_1 \\ x_2 \end{pmatrix} + \begin{pmatrix} \frac{1}{L} \\ 0 \end{pmatrix} u \,.$$

If the angle speed ω is chosen as the output, it leaves:

$$y = x_2 = \begin{pmatrix} 0 & 1 \end{pmatrix} \begin{pmatrix} x_1 \\ x_2 \end{pmatrix}$$

If the angle θ is chosen as the output, then the above two state variables are not enough to represent the dynamics of the system, and another state variable, x_3, should be introduced:

$$x_3 = \theta \,.$$

So

$$\dot{x}_3 = \dot{\theta} = x_2 \,.$$

The state equation is:

$$\begin{pmatrix} \dot{x}_1 \\ \dot{x}_2 \\ \dot{x}_3 \end{pmatrix} = \begin{pmatrix} -\frac{R}{L} & -\frac{K_b}{L} & 0 \\ \frac{K_a}{J} & -\frac{B}{J} & 0 \\ 0 & 1 & 0 \end{pmatrix} \begin{pmatrix} x_1 \\ x_2 \\ x_3 \end{pmatrix} + \begin{pmatrix} \frac{1}{L} \\ 0 \\ 0 \end{pmatrix} u \,. \tag{1.28}$$

The output equation is:

$$y = x_3 = \begin{pmatrix} 0 & 0 & 1 \end{pmatrix} \begin{pmatrix} x_1 \\ x_2 \\ x_3 \end{pmatrix} \,.$$

The Input-Output Model

When a system is created by the input-output model, i.e., transfer function or differential equation, the state space model can be derived with the input-output model. This process is called realization. Based on different types of the input-output model, different algorithms can be adopted for realization. The detailed procedure can be found in Section 1.3.5.

1.3 Transition From One Mathematical Model to Another

As we know, every mathematical model has advantages and disadvantages. To use the advantages of all mathematical models, one must have the flexibility of transition from one model to another. This issue of transition is, obviously, of great practical and theoretical importance. In the following, we present some transition methods.

1.3.1 From Differential Equation to Transfer Function for Single-input-single-output Systems

Case 1. The Righthand Side of the Differential Equation Does Not Involve Derivatives
Consider a single-input-single-output (SISO) system described by the following differential equation:

$$y^{(n)} + \alpha_{n-1}y^{(n-1)} + \cdots + \alpha_1 y^{(1)} + \alpha_0 y = \beta_0 u , \tag{1.29}$$

where all the system's initial conditions are assumed to be zero, i.e., $y^{(k)}(0) = 0$, for $k = 1, 2, \ldots, n-1$. Applying the Laplace transform to equation (1.29) can result in:

$$s^n Y(s) + \alpha_{n-1}s^{n-1} Y(s) + \cdots + \alpha_1 s Y(s) + \alpha_0 Y(s) = \beta_0 U(s) .$$

Hence, the transfer function is given by:

$$G(s) = \frac{Y(s)}{U(s)} = \frac{\beta_0}{s^n + \alpha_{n-1}s^{n-1} + \cdots + \alpha_1 s + \alpha_0} . \tag{1.30}$$

Case 2. The Righthand Side of the Differential Equation Involves Derivatives
Consider a SISO system described by the differential equation:

$$y^{(n)} + \alpha_{n-1}y^{(n-1)} + \cdots + \alpha_1 y^{(1)} + \alpha_0 y = \beta_m u^{(m)} + \cdots + \beta_1 u^{(1)} + \beta_0 u , \tag{1.31}$$

where $m < n$ and all initial conditions are assumed to be zero, i.e., $y^{(k)}(0) = 0$, for $k = 0, 1, \ldots, n-1$. We can determine the transfer equation (1.31) as follows: suppose $z(t)$ is the solution of equation (1.29), with $\beta_0 = 1$. When the superposition principle is used, the solution $y(t)$ of equation (1.31) will be obtained:

$$y(t) = \beta_m z^{(m)} + \beta_{m-1}z^{(m-1)} + \cdots + \beta_1 z^{(1)} + \beta_0 z . \tag{1.32}$$

By applying the Laplace transformation to equation (1.32), we obtain:

$$Y(s) = \beta_m s^m Z(s) + \beta_{m-1}s^{m-1} Z(s) + \cdots + \beta_1 s Z(s) + \beta_0 Z(s) . \tag{1.33}$$

Here, we have set $z^{(k)}(0) = 0$, for $k = 0, 1, \ldots, n-1$. When $\beta_0 = 1$, the solution of equation (1.29) is $z(t) = y(t)$. Here, it is assumed that all initial conditions of $y(t)$, hence, of $z(t)$, are zero. In equation (1.30), when $\beta_0 = 1$, we have:

$$Z(s) = \left[\frac{1}{s^n + \alpha_{n-1} s^{n-1} + \cdots + \alpha_1 s + \alpha_0} \right] U(s) . \tag{1.34}$$

By substituting equation (1.34) into (1.33), the transfer function $G(s)$ of the differential equation (1.31) is obtained as:

$$G(s) = \frac{Y(s)}{U(s)} = \frac{\beta_m s^m + \beta_{m-1} s^{m-1} + \cdots + \beta_1 s + \beta_0}{s^n + \alpha_{n-1} s^{n-1} + \cdots + \alpha_1 s + \alpha_0} . \tag{1.35}$$

Remark 1.3.1. The transfer function $G(s)$, given by equation (1.35), can be easily derived from equation (1.31) if we set s^k in place of the kth derivative and replace $y(t)$ and $u(t)$ with $Y(s)$ and $U(s)$, respectively. That is, we can derive equation (1.35) by replacing $y^{(k)}(t)$ with $s^k Y(s)$, and $u^{(k)}(t)$ with $s^k U(s)$ in equation (1.31).

1.3.2 From Transfer Function to Differential Equation for SISO Systems

Suppose a SISO system is described by equation (1.35). Then, working backwards using the method given in Remark 1.3.1, the differential equation (1.31) can be constructed by substituting s^k with the kth derivative and $Y(s)$ and $U(s)$ with $y(t)$ and $u(t)$, respectively.

1.3.3 From $G(s)$ to $g(t)$ and Vice Versa

The matrices $G(s)$ and $g(t)$ are related through the Laplace transform:

$$L\{g(t)\} = G(s) \quad \text{or} \quad g(t) = L^{-1}\{G(s)\} . \tag{1.36}$$

1.3.4 From State Equations to Transfer Function Matrix

Consider a system described by the following state equations:

$$\dot{x}(t) = Ax(t) + Bu(t)$$
$$y(t) = Cx(t) + Du(t) \tag{1.37}$$
$$x(t_0) = x(0) = x_0 .$$

Take the Laplace transformation to both sides of equation (1.37):

$$sx(s) - x(0) = Ax(s) + Bu(s)$$
$$y(s) = Cx(s) + Du(s)$$
$$x(s) = (sI - A)^{-1} x(0) + (sI - A)^{-1} Bu(s)$$
$$y(s) = C(sI - A)^{-1} x(0) + C(sI - A)^{-1} Bu(s) + Du(s) .$$

For zero initial condition, we have:

$$y(s) = \left[C(sI - A)^{-1}B + D \right] u(s) .$$

Then the system's transfer function matrix $G(s)$ is given by the relation:

$$G(s) = C(sI - A)^{-1}B + D . \qquad (1.38)$$

Example 1.12. Derive the transfer function of the following state space equation.

$$\dot{x} = \begin{pmatrix} 0 & 0 & 1 & 0 \\ 0 & 0 & 0 & 1 \\ 0 & 0 & -1 & 0 \\ 0 & 0 & 0 & -1 \end{pmatrix} x + \begin{pmatrix} 0 & 1 \\ 1 & 1 \\ 1 & 0 \\ 0 & -2 \end{pmatrix} u ,$$

$$y = \begin{pmatrix} 1 & 0 & 0 & 0 \\ 0 & 1 & 0 & 0 \end{pmatrix} x .$$

Solution.

$$A = \begin{pmatrix} 0 & 0 & 1 & 0 \\ 0 & 0 & 0 & 1 \\ 0 & 0 & -1 & 0 \\ 0 & 0 & 0 & -1 \end{pmatrix}, \quad B = \begin{pmatrix} 0 & 1 \\ 1 & 1 \\ 1 & 0 \\ 0 & -2 \end{pmatrix}, \quad C = \begin{pmatrix} 1 & 0 & 0 & 0 \\ 0 & 1 & 0 & 0 \end{pmatrix}, \quad D = 0 ,$$

$$(sI - A)^{-1} = \begin{pmatrix} \frac{1}{s} & 0 & \frac{1}{s(s+1)} & 0 \\ 0 & \frac{1}{s} & 0 & \frac{1}{s(s+1)} \\ 0 & 0 & \frac{1}{s+1} & 0 \\ 0 & 0 & 0 & \frac{1}{s+1} \end{pmatrix} .$$

The result can be obtained, according to (1.38):

$$G(s) = \begin{bmatrix} \frac{1}{s(s+1)} & \frac{1}{s} \\ \frac{1}{s} & \frac{s-1}{s(s+1)} \end{bmatrix} .$$

MATLAB can be adopted to compute this equation. Type:

```
a=[0 0 1 0;0 0 0 1;0 0 -1 0;0 0 0 -1];
b=[0 1;1 1;1 0;0 -2];
c=[1 0 0 0;0 1 0 0];
d=[0 0;0 0];
[N1,d1]=ss2tf(a,b,c,d,1)
[N2,d2]=ss2tf(a,b,c,d,2)
```

which yields

```
N1 =
    0       0        1.0000      1.0000      0.0000
    0    1.0000      2.0000      1.0000      0
d1 =
    1    2    1    0    0
N2 =
    0    1.0000      2.0000      1.0000      0
    0    1.0000      0.0000     -1.0000      0.0000
d2 =
    1    2    1    0    0
```

Thus, the transfer matrix is:

$$G(s) = \begin{bmatrix} \frac{s^2+s}{s^4+2s^3+s^2} & \frac{s^3+2s^2+s}{s^4+2s^3+s^2} \\ \frac{s^3+2s^2+s}{s^4+2s^3+s^2} & \frac{s^3-s}{s^4+2s^3+s^2} \end{bmatrix}.$$

Simplifying yields:

$$G(s) = \begin{bmatrix} \frac{1}{s(s+1)} & \frac{1}{s} \\ \frac{1}{s} & \frac{s-1}{s(s+1)} \end{bmatrix}.$$

Here, [N1,d1]=ss2tf(a,b,c,d,1) computes the transfer matrix from the first input to all outputs, e.g., the first column of $G(s)$. N1 is the numerator coefficient of the first column of $G(s)$, d1 is the denominator coefficient of the first column of $G(s)$. In a similar way, [N2,d2]=ss2tf(a,b,c,d,2) computes the transfer matrix from the second input to all outputs.

1.3.5 From Transfer Function Matrix to State Equations for SISO Systems

The transition from $G(s)$ to state equations is the well known problem of the state space realization. This is, in general, a difficult problem and has been (and still remains) a topic to study. In the following, we will present some introductory results regarding this problem.

Imagine a system is described by a scalar transfer function with the form:

$$G(s) = \frac{Y(s)}{U(s)} = \frac{\beta_{n-1}s^{n-1} + \beta_{n-1}s^{n-1} + \cdots + \beta_1 s + \beta_0}{s^n + \alpha_{n-1}s^{n-1} + \cdots + \alpha_1 s + \alpha_0}, \tag{1.39}$$

or, equivalently, by the differential equation:

$$y^{(n)} + \alpha_{n-1}y^{(n-1)} + \cdots + \alpha_1 y^{(1)} + \alpha_0 y = \beta_n u^{(n)} + \beta_{n-1} u^{(n-1)} \cdots + \beta_1 u^{(1)} + \beta_0 u. \tag{1.40}$$

Equation (1.40) can be expressed in the form of two equations as follows:

$$z^{(n)} + \alpha_{n-1}z^{(n-1)} + \cdots + \alpha_1 z^{(1)} + \alpha_0 z = u, \tag{1.41}$$

$$y(t) = \beta_{n-1}z^{(n-1)} + \cdots + \beta_1 z^{(1)} + \beta_0 z. \tag{1.42}$$

Suppose $z(t)$ is the solution of equation (1.41). Then, the solution of equation (1.40) will be given by equation (1.42).

The state variables $x_1, x_2, \ldots, x_n$ are defined as follows:

$$
\begin{aligned}
x_1(t) &= z(t) , \\
x_2(t) &= z^{(1)}(t) = x_1^{(1)}(t) , \\
x_3(t) &= z^{(2)}(t) = x_2^{(1)}(t) , \\
&\vdots \\
x_n(t) &= z^{(n-1)}(t) = x_{n-1}^{(1)}(t) .
\end{aligned}
\tag{1.43}
$$

Substituting equation (1.42) into equation (1.41) can result in:

$$
\dot{x}_n(t) = -\alpha_{n-1} x_n(t) - \cdots - \alpha_1 x_2(t) - \alpha_0 x_1(t) + u(t) .
\tag{1.44}
$$

Also, substituting equations (1.43) into equation (1.42) can result in:

$$
y(t) = \beta_{n-1} x_n(t) + \beta_{n-2} x_{n-1}(t) + \cdots + \beta_1 x_2(t) + \beta_0 x_1(t) .
\tag{1.45}
$$

Equations (1.43) to (1.45) can be expressed in a matrix form:

$$
\begin{aligned}
\dot{x}(t) &= Ax(t) + bu(t) \\
y(t) &= c^{T} x(t) ,
\end{aligned}
\tag{1.46}
$$

where $x^{T} = (x_1, x_2, \ldots, x_n)$ and:

$$
A = \begin{bmatrix}
0 & 1 & 0 & 0 & \cdots & 0 \\
0 & 0 & 1 & 0 & \cdots & 0 \\
\vdots & \vdots & \vdots & \vdots & & \vdots \\
0 & 0 & 0 & 0 & \cdots & 1 \\
-\alpha_0 & -\alpha_1 & -\alpha_2 & -\alpha_3 & \cdots & -\alpha_{n-1}
\end{bmatrix} ,
$$

$$
b = \begin{bmatrix}
0 \\
0 \\
\vdots \\
0 \\
1
\end{bmatrix} ,
\tag{1.47}
$$

$$
c = \begin{bmatrix}
\beta_0 \\
\beta_1 \\
\vdots \\
\beta_{n-2} \\
\beta_{n-1}
\end{bmatrix} .
$$

Hence, equations (1.46) constitute the state equations' description of the transfer function (1.39).

Due to the special form of matrix A and vector b, we say that the state equations (1.46) are in *phase canonical form*, while the state variables are called *phase variables*. Phase variables are, in general, state variables, which are defined according to equations (1.43); i.e., every state variable is the derivative of the previous one. In particular, the special form of matrix A and vector b is characterized as follows:

If the first column and the last row in matrix A are deleted, then a $(n-1) \times (n-1)$ unit matrix is revealed. Also, the elements of the last row of A are the coefficients of the differential equation (1.40), placed in reverse order and all with negative signs. The vector b has all its elements equal to zero, except for the nth element, which is equal to one.

Example 1.13. The differential equation mathematical model of a system is given as:

$$\dddot{y} + 6\ddot{y} + 41\dot{y} + 7y = 6u \,.$$

Try to derive the state equation and the output equation.

Solution. We choose $y/6$, $\dot{y}/6$, $\ddot{y}/6$ as the state variables:

$$x_1 = \frac{y}{6}\,, \quad x_2 = \frac{\dot{y}}{6}\,, \quad x_3 = \frac{\ddot{y}}{6}\,.$$

Then,

$$\dot{x}_1 = \frac{\dot{y}}{6} = x_2\,,$$

$$\dot{x}_2 = \frac{\ddot{y}}{6} = x_3\,,$$

$$\dot{x}_3 = \frac{\dddot{y}}{6} = -7x_1 - 41x_2 - 6x_3 + u\,.$$

The state equation is:

$$\begin{pmatrix} \dot{x}_1 \\ \dot{x}_2 \\ \dot{x}_3 \end{pmatrix} = \begin{pmatrix} 0 & 1 & 0 \\ 0 & 0 & 1 \\ -7 & -41 & -6 \end{pmatrix} \begin{pmatrix} x_1 \\ x_2 \\ x_3 \end{pmatrix} + \begin{pmatrix} 0 \\ 0 \\ 1 \end{pmatrix} u \,. \qquad (1.48)$$

The output equation is:

$$y = 6x_1 = \begin{pmatrix} 6 & 0 & 0 \end{pmatrix} \begin{pmatrix} x_1 \\ x_2 \\ x_3 \end{pmatrix} \,.$$

1.4 Summary

Three kinds of models and their relationship were introduced in this chapter. Each type of model has its merits and shortages. The transfer function is widely used in classical control theory, which mainly studies the SISO linear system in Laplace domain. The differential equation is used in time domain, and the state space equation is commonly used in modern control theory, in which the MIMO system is studied in time domain. The following contents of modern control theory, which studies the properties of a system and system synthesis, are mainly based on the state space description.

Appendix: Three Power Generation Models

Case 1) Thermal Power Generation

The fundamental dynamics of a 160 MW drum type boiler, turbine, generator (BTG) plant can be represented by a third order MIMO coupling nonlinear model over a wide operating range. Typically, the coordinated control governs the dominant behavior of the power unit through the power and steam pressure control loops, as shown in Figure 1.13.

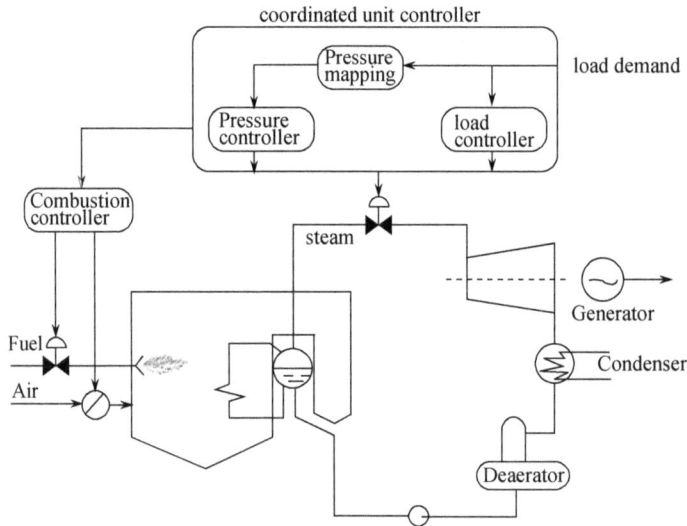

Fig. 1.13: Coordinated control scheme.

The inputs are positions of valve actuators that control the mass flow rates of fuel (u_1 in pu), steam to the turbine (u_2 in pu), and feedwater to the drum (u_3 in pu). The outputs are electric power (E in MW), drum steam pressure (P in kg/cm^2), and drum water level deviation (L in mm). The state variables are electric power (E), drum steam pressure (P), and fluid (steam water) density (ρ_f). The state equations are:

$$\frac{dP}{dt} = 0.9u_1 - 0.0018u_2 P^{9/8} - 0.15u_3 ,$$

$$\frac{dE}{dt} = ((0.73u_2 - 0.16)P^{9/8} - E)/10 , \tag{1.49}$$

$$\frac{d\rho_f}{dt} = (141u_3 - (1.1u_2 - 0.19)P)/85 .$$

The drum water level output is calculated using the following algebraic equations:

$$q_e = (0.85u_2 - 0.14)P - 45.59u_1 - 2.51u_3 - 2.09 ,$$

$$\alpha_s = (1/\rho_f - 0.0015)/(1/(0.8P - 25.6) - 0.0015) , \tag{1.50}$$

$$L = 50(0.13\rho_f + 60\alpha_s + 0.11q_e - 65.5) ,$$

where α_s is the steam quality and q_e is the evaporation rate (kg/s).

Define x_1, x_2 and x_3 as the drum steam pressure (kg/cm^2), the electric power (MW), and the steam water fluid density in the drum, respectively. The output y_3 is the drum water level (cm) calculated using two algebraic calculations, α_{cs} and q_e, which are the steam quality and the evaporation rates (kg/s).The input u_1, u_2, and u_3 are normalized positions of valve actuators that control the mass flow rates of fuel, steam to the turbine, and feedwater to the drum, respectively. Then the state space equation becomes:

$$\dot{x}_1 = -0.0018u_2 x_1^{9/8} + 0.9u_1 - 0.15u_3$$

$$\dot{x}_2 = (0.073u_2 - 0.016) x_1^{9/8} - 0.1x_2$$

$$\dot{x}_3 = [141u_3 - (1.1u_2 - 0.19) x_1]/85$$

$$y_1 = x_1$$

$$y_2 = x_2 \tag{1.51}$$

$$y_3 = 0.05 (0.13073x_3 + 100\alpha_{cs} + q_e/9 - 67.975)$$

$$\alpha_{cs} = \frac{(1 - 0.001538x_3)(0.8x_1 - 25.6)}{x_3 (1.0394 - 0.0012304x_1)}$$

$$q_e = (0.845u_2 - 0.147) x_1 + 45.59u_1 - 2.514u_3 - 2.096 .$$

Case 2) Nuclear Power Generation

The water level model developed by E. Irving is a fourth-order transfer function representation, with its power level dependent parameters listed in Table 1:

$$Y(s) = \frac{G_1}{s}(Q_w(s) - Q_v(s)) - \frac{G_2}{1 + \tau_2 s}(Q_w(s) - Q_v(s)) + \frac{G_3 s}{\tau_1^{-2} + 4\pi^2 T^{-2} + 2\tau_1^{-2} s + s^2} Q_w(s),$$

$$\tag{1.52}$$

Tab. 1: Parameters of the Steam Generator Model With Respect to Operating Power.

P (% power)	5	15	30	50	100
q_v (kg/s)	57.4	180.8	381.7	660	1435
G_1	0.058	0.058	0.058	0.058	0.058
G_2	9.63	4.46	1.83	1.05	0.47
G_3	0.181	0.226	0.310	0.215	0.105
τ_1	41.9	26.3	43.4	34.8	28.6
τ_2	48.4	21.5	4.5	3.6	3.4
T	119.6	60.5	17.7	14.2	11.7

where $Y(s)$, $Q_w(s)$ and $Q_v(s)$ represent the water level, the feed water flow rate and the steam flow rate respectively. τ_1, τ_2 and T are the damping time constants and the oscillation period.

Suppose that:

$$Y_1(s) = \frac{G_1}{s}(Q_w(s) - Q_v(s)),$$

$$Y_2(s) = -\frac{G_2}{1 + \tau_2 s}(Q_w(s) - Q_v(s)),$$

$$Y_3(s) = \frac{G_3 s}{\tau_1^{-2} + 4\pi^2 T^{-2} + 2\tau_1^{-1}s + s^2}Q_w(s).$$

(1.53)

Define the state variables as shown in the Figure 1.14.

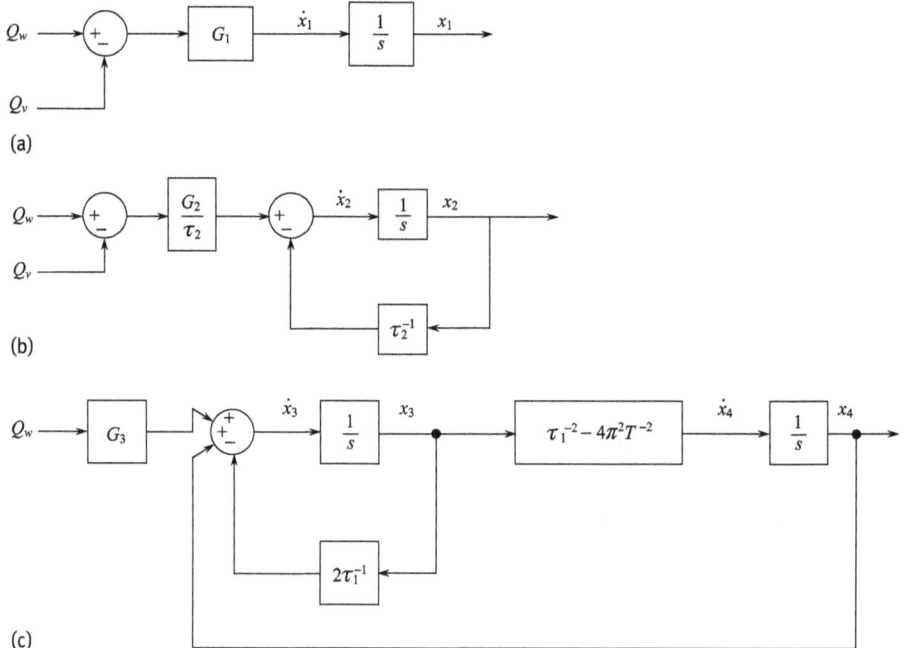

(a)

(b)

(c)

Fig. 1.14: (a) $Y_1(s)$; (b) $Y_2(s)$; (c) $Y_3(s)$.

Then,

$$\dot{x}_1(t) = G_1(Q_w(t) - Q_v(t))$$

$$\dot{x}_2(t) = -\tau_2^{-1}x_2(t) - \frac{G_2}{\tau_2}(Q_w(t) - Q_v(t))$$

$$\dot{x}_3(t) = -2\tau_1^{-1}x_3(t) + x_4(t) + G_3 Q_w(t)$$

$$\dot{x}_4(t) = -(\tau_1^{-2} + 4\pi^2 T^{-2})x_3(t) .$$

(1.54)

The control variable $u = Q_w$, the disturbance $d = Q_v$, and the water level output $y = x_1 + x_2 + x_3$, and equations (1.54) can be rearranged in the following state space equations:

$$\begin{cases} \dot{x}_c(t) = A(\theta)x_c(t) + B(\theta)u_c(t) + W(\theta)d_c(t) \\ y_c(t) = Cx_c(t) , \end{cases}$$

(1.55)

where

$$x_c = \begin{bmatrix} x_1 \\ x_2 \\ x_3 \\ x_4 \end{bmatrix} ,$$

$$A(\theta) = \begin{bmatrix} 0 & 0 & 0 & 0 \\ 0 & a_{22} & 0 & 0 \\ 0 & 0 & a_{33} & a_{34} \\ 0 & 0 & a_{43} & 0 \end{bmatrix} , \quad B(\theta) = \begin{bmatrix} b_1 \\ b_2 \\ b_3 \\ 0 \end{bmatrix} , \quad W(\theta) = \begin{bmatrix} d_1 \\ d_2 \\ 0 \\ 0 \end{bmatrix} ,$$

$$C = \begin{bmatrix} 1 & 1 & 1 & 0 \end{bmatrix}$$

$$a_{22} = -\tau_2^{-1} , \quad a_{33} = -2\tau_1^{-1} , \quad a_{34} = 1 , \quad a_{43} = -(\tau_1^{-1} + 4\pi^2 T^{-2})$$

$$b_1 = G_1 , \quad b_2 = -G_2\tau_2^{-1} , \quad b_3 = G_3$$

$$d_1 = -G_1 , \quad d_2 = \frac{G_2}{\tau_2} .$$

Case 3) Wind Power Generation

The doubly fed induction generators (DFIGs) have been widely used in the modern wind energy systems, due to the advantages of variable speed operation and four quadrant active and reactive power capabilities. In DFIG, the stator is directly connected to the power grid, while the rotor is connected to the grid by a bidirectional converter, as shown in Figure 1.15. This converter controls active and the reactive power between the stator and ac supply, or a standalone grid. The equivalent circuit of a DFIG in an arbitrary reference frame rotating at synchronous angular speed ω_1 is shown in Figure 1.16.

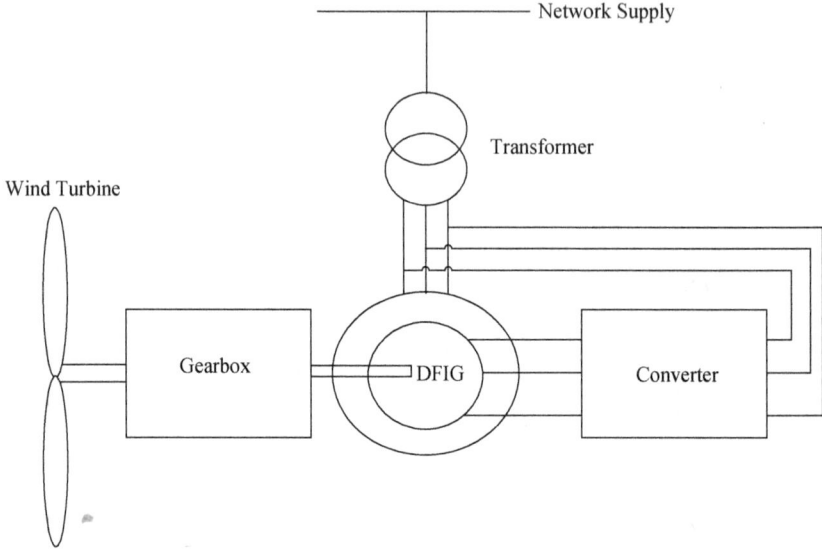

Fig. 1.15: The doubly fed induction generator (DFIG) system.

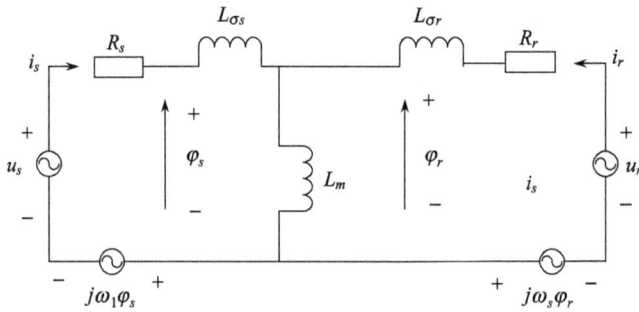

Fig. 1.16: The equivalent circuit of a DFIG.

The DFIG model in the synchronous reference frame is given by:

$$u_s = R_s i_s + d\varphi_s/dt + j\omega_1\varphi_s \,, \tag{1.56}$$
$$u_r = R_r i_r + d\varphi_r/dt + j(\omega_1 - \omega_r)\,\varphi_r \,, \tag{1.57}$$

where the relationship between fluxes and currents is:

$$\varphi_s = L_s i_s + L_m i_r \,, \tag{1.58}$$
$$\varphi_r = L_m i_s + L_r i_r \,, \tag{1.59}$$

and generator active and reactive power are:

$$P = \frac{3}{2}\left(u_{sd}i_{sd} + u_{sq}i_{sq}\right), \tag{1.60}$$

$$Q = \frac{3}{2}\left(u_{sq}i_{sd} + u_{sd}i_{sq}\right). \tag{1.61}$$

The subscripts s and r represent the stator and rotor parameters respectively ω_1 represents the synchronous speed and ω_r represents rotor angular speed. R_s and R_r represent stator and the rotor windings per phase electrical resistance and L_s, L_r and L_m represent the proper and the mutual inductances of the stator and rotor windings. u represents voltage vector.

The DFIG power control aims independent stator active P and reactive power Q control by means of a rotor current regulation. For this purpose, P and Q are represented as functions of each individual rotor current. We use stator flux oriented control, that decouples the dq axis, which means: $\varphi_{sd} = \varphi_s$, $\varphi_{sq} = 0$. Thus, (1.58) becomes:

$$i_{sd} = \frac{\varphi_s}{L_s} - \frac{L_m}{L_s}i_{rd}, \tag{1.62}$$

$$i_{sq} = -\frac{L_m}{L_s}i_{rq}. \tag{1.63}$$

Similarly, using stator flux orientation, the stator voltage becomes $u_{sd} = 0$ and $u_{sq} = u_s$. Hence, the active (1.60) and reactive (1.61) power can be calculated by using (1.62) and (1.63):

$$P = -\frac{3}{2}u_s\frac{L_m}{L_s}i_{rq}, \tag{1.64}$$

$$Q = \frac{3}{2}\left(\frac{\varphi_s}{L_s} - \frac{L_m}{L_s}i_{rd}\right)u_s. \tag{1.65}$$

Thus, rotor currents will reflect on stator current and on stator active and reactive power, respectively. Consequently, this principle can be used on stator active and reactive power control of the DFIG.

The DFIG power control is realized by the rotor currents control using (1.64) and (1.65). Using (1.62) and (1.63), the rotor voltage (1.57), in the synchronous referential frame, becomes:

$$u_r = (R_r + j\sigma L_r\omega_{sl})\,i_r + \sigma L_r\frac{di_r}{dt} + j\frac{L_m}{L_s}\omega_{sl}\varphi_s, \tag{1.66}$$

then:

$$\frac{di_r}{dt} = \frac{-R_r}{\sigma L_r}i_r - j\omega_{sl}i_r + \frac{1}{\sigma L_r}u_r - j\frac{L_m}{\sigma L_rL_s}\omega_{sl}\varphi_s, \tag{1.67}$$

$$\frac{di_r}{dt} = -\frac{R_r}{\sigma L_r}i_{rd} + \omega_{sl}i_{rq} + \frac{1}{\sigma L_r}u_{rd}$$

$$\frac{di_r}{dt} = -\omega_{sl}i_{rd} - \frac{R_r}{\sigma L_r}i_{rq} + \frac{1}{\sigma L_r}u_{rq} - \frac{\omega_{sl}L_m}{\sigma L_rL_s}\varphi_s. \tag{1.68}$$

(1.68) can be expressed as:

$$
\begin{bmatrix} \frac{di_{rd}}{dt} \\ \frac{di_{rq}}{dt} \end{bmatrix} = \begin{bmatrix} -\frac{R_r}{\sigma L_r} & \omega_{sl} \\ -\omega_{sl} & -\frac{R_r}{\sigma L_r} \end{bmatrix} \begin{bmatrix} i_{rd} \\ i_{rq} \end{bmatrix} + \begin{bmatrix} \frac{1}{\sigma L_r} & 0 \\ 0 & \frac{1}{\sigma L_r} \end{bmatrix} \begin{bmatrix} u_{rd} \\ u_{rq} \end{bmatrix} + \begin{bmatrix} 1 & 0 \\ 0 & 1 \end{bmatrix} \begin{bmatrix} 0 \\ -\frac{\omega_{sl} L_m}{\sigma L_r L_s} \varphi_s \end{bmatrix}
$$

$$
\begin{bmatrix} i_{rd} \\ i_{rq} \end{bmatrix} = \begin{bmatrix} 1 & 0 \\ 0 & 1 \end{bmatrix} \begin{bmatrix} i_{rd} \\ i_{rq} \end{bmatrix},
$$

(1.69)

where

$$
\omega_{sl} = \omega_1 - \omega_r \quad \text{and} \quad \sigma = 1 - \frac{L_m^2}{L_s L_r}.
$$

Assuming that:

$$
x = [x_1, x_2]^T = [i_{rd}, i_{rq}]^T, \qquad \dot{x} = [\dot{x}_1, \dot{x}_2]^T = [\dot{i}_{rd}, \dot{i}_{rq}]^T,
$$
$$
u = [u_1, u_2]^T = [u_{rd}, u_{rq}]^T, \qquad y = [y_1, y_2]^T = [i_{rd}, i_{rq}]^T,
$$

(1.69) can be expressed in the standard space state form:

$$
\dot{x} = Ax + Bu + G\omega
$$
$$
y = Cx,
$$

(1.70)

where:

$$
A = \begin{bmatrix} -\frac{R_r}{\sigma L_r} & \omega_{sl} \\ -\omega_{sl} & -\frac{R_r}{\sigma L_r} \end{bmatrix}, \quad B = \begin{bmatrix} \frac{1}{\sigma L_r} & 0 \\ 0 & \frac{1}{\sigma L_r} \end{bmatrix},
$$
$$
C = \begin{bmatrix} 1 & 0 \\ 0 & 1 \end{bmatrix}, \quad G\omega = \begin{bmatrix} 0 \\ -\frac{\omega_{sl} L_m}{\sigma L_r L_s} \varphi_s \end{bmatrix}.
$$

Exercise

1.1. Consider the network shown in Figure 1.17. Derive the network's differential equation mathematical model.

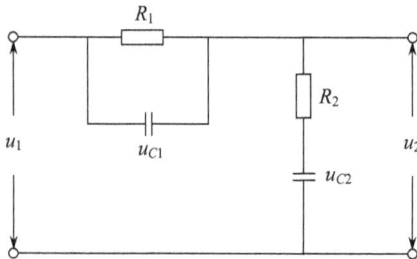

Fig. 1.17: Circuit diagram.

1.2. Consider the mechanical system shown in Figure 1.18, where $x_1 = u_{C1}$, $x_2 = u_{C2}$, K, B, M are the spring's constant, the friction coefficient and the mass, respectively. Derive the state space equations.

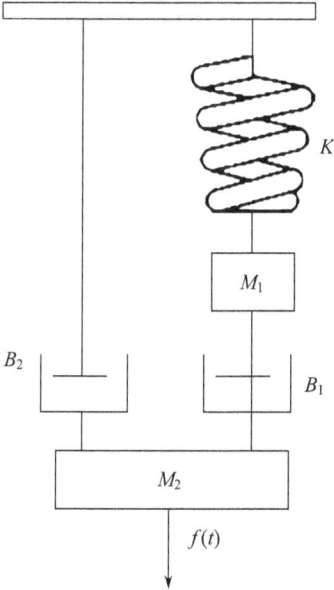

Fig. 1.18: Mechanical system.

1.3. Consider the network shown in Figure 1.17. Derive the network's state space equations.

1.4. Consider the network shown in Figure 1.19. Derive the network's transfer function

$$\frac{U_C(s)}{U(s)}.$$

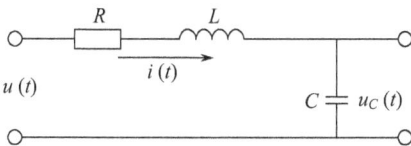

Fig. 1.19: Network.

1.5. Consider the network shown in Figure 1.19. Derive the state space equations.

1.6. Consider the network shown in Figure 1.20. Derive the network's transfer function

$$\frac{U_L(s)}{U_1(s)}, \quad \frac{U_L(s)}{U_2(s)}.$$

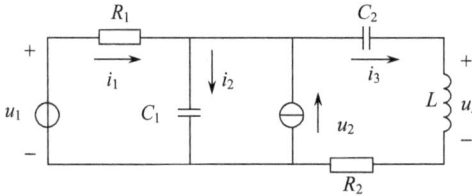

Fig. 1.20: Network.

1.7. Consider the network shown in Figure 1.20. Derive the network's state space equations.

1.8. Transition from state equations to transfer function.

$$(1) \quad \begin{bmatrix} \dot{x}_1 \\ \dot{x}_2 \\ \dot{x}_3 \end{bmatrix} = \begin{bmatrix} 0 & 0 & 0 \\ 1 & -2 & 0 \\ 0 & 1 & -3 \end{bmatrix} \begin{bmatrix} x_1 \\ x_2 \\ x_3 \end{bmatrix} + \begin{bmatrix} 6 \\ 0 \\ 0 \end{bmatrix} u , \quad (2) \quad \begin{bmatrix} \dot{x}_1 \\ \dot{x}_2 \end{bmatrix} = \begin{bmatrix} -5 & 0 \\ 1 & -1 \end{bmatrix} \begin{bmatrix} x_1 \\ x_2 \end{bmatrix} + \begin{bmatrix} 2 \\ 0 \end{bmatrix} u ,$$

$$y = \begin{bmatrix} 0 & 1 & -2 \end{bmatrix} \begin{bmatrix} x_1 \\ x_2 \\ x_3 \end{bmatrix} . \qquad\qquad y = \begin{bmatrix} 1 & -\frac{1}{2} \end{bmatrix} \begin{bmatrix} x_1 \\ x_2 \end{bmatrix} .$$

1.9. Transition from transfer function to state equations.

$$(1) \quad g(s) = \frac{s^3 + s + 1}{s^3 + 6s^2 + 11s + 6} .$$

$$(2) \quad g(s) = \frac{s^2 + 2s + 3}{s^3 + 2s^2 + 3s + 1} .$$

1.10. Transition from differential equation to transfer function.

$$(1) \quad 2\ddot{y} + 3\dot{y} = \ddot{u} - u .$$
$$(2) \quad \dddot{y} + 2\ddot{y} + 3\dot{y} + 5y = 5\ddot{u} + 7u .$$

1.11. Transition from transfer function to differential equation.

$$g(s) = \frac{Y(s)}{U(s)} = \frac{160s + 720}{s^3 + 16s^2 + 194s + 640} .$$

2 Linear Transformation of State Vector

2.1 Linear Algebra

This chapter reviews a number of concepts and results in linear algebra that are essential in the linear transformation of state vector.

As we saw in the preceding chapter, all parameters that arise in the real world are real numbers. Therefore, we deal only with real numbers throughout this chapter, except when specifically stated otherwise. Suppose A, B, C, and D are, respectively, $n \times m$, $m \times r$, $l \times n$, and $r \times p$ are dimensional real matrices. Let a_i be the ith column of A, and b_j the jth row of B. Then we have:

$$AB = \begin{bmatrix} a_1 & a_2 & \cdots & a_m \end{bmatrix} \begin{bmatrix} b_1 \\ b_2 \\ \vdots \\ b_m \end{bmatrix} = a_1 b_1 + a_2 b_2 + \cdots + a_m b_m , \qquad (2.1)$$

$$CA = C \begin{bmatrix} a_1 & a_2 & \cdots & a_m \end{bmatrix} = \begin{bmatrix} Ca_1 & Ca_2 & \cdots & Ca_m \end{bmatrix} , \qquad (2.2)$$

$$BD = \begin{bmatrix} b_1 \\ b_2 \\ \vdots \\ b_m \end{bmatrix} , \qquad D = \begin{bmatrix} b_1 D \\ b_2 D \\ \vdots \\ b_m D \end{bmatrix} . \qquad (2.3)$$

These identities can easily be verified. Note that $a_i b_i$ is a $n \times r$ matrix; it is the product of a $n \times 1$ column vector and a $1 \times r$ row vector. The product $b_i a_i$ is not defined unless $n = r$; it becomes a scalar if $n = r$.

Consider a n-dimensional real linear space, denoted by R^n. Every vector in R^n has n real numbers such as:

$$x = \begin{bmatrix} x_1 \\ x_2 \\ \vdots \\ x_n \end{bmatrix} .$$

To save space, we write it as $x = [x_1 \ x_2 \ \cdots \ x_n]^T$, where the superscript denotes the transpose.

The set of vectors $\{x_1, x_2, \ldots, x_m\}$ in R^n is said to be linearly dependent if there exists set of real numbers $\alpha_1, \alpha_2, \ldots, \alpha_m$, which are not all zero, such that:

$$\alpha_1 x_1 + \alpha_2 x_2 + \cdots + \alpha_m x_m = 0 . \qquad (2.4)$$

If the only set of α_i that makes (2.4) hold is $\alpha_1 = 0$, $\alpha_2 = 0$, ..., $\alpha_m = 0$, then the set of vectors $\{x_1, x_2, \ldots, x_m\}$ is said to be linearly independent.

https://doi.org/10.1515/9783110574951-002

If the set of vectors in (2.4) is linearly dependent, then there exists at least one a_i, such as a_1, that is nonzero. Then (2.4) implies:

$$x_1 = -\frac{1}{\alpha_1} [\alpha_2 x_2 + \alpha_3 x_3 + \cdots + \alpha_m x_m] = \beta_2 x_2 + \beta_3 x_3 + \cdots + \beta_m x_m ,$$

where $\beta_i = -\alpha_i/\alpha_1$. Such an expression is called a linear combination.

The dimension of a linear space can be defined as the maximum number of linearly independent vectors in the space. Thus, in R^n, we can find n linearly independent vectors.

Basis and Representation

A set of linearly independent vectors in R^n is called a basis if every vector in R^n can be expressed as a unique linear combination of the set. In R^n, any set of n linearly independent vectors can be used as a basis. Let $\{q_1, q_2, \ldots, q_n\}$ be such a set. Then every vector x can be expressed uniquely as:

$$x = \alpha_1 q_1 + \alpha_2 q_2 + \cdots + \alpha_n q_n . \tag{2.5}$$

Define the $n \times n$ square matrix:

$$Q = \begin{bmatrix} q_1 & q_2 & \cdots & q_n \end{bmatrix} . \tag{2.6}$$

Then (2.5) can be written as:

$$x = Q \begin{bmatrix} \alpha_1 \\ \alpha_2 \\ \vdots \\ \alpha_n \end{bmatrix} = Q\bar{x} . \tag{2.7}$$

We call $\bar{x} = [\alpha_1 \ \alpha_2 \ \cdots \ \alpha_n]^\mathsf{T}$ the representation of the vector x, with respect to the basis $\{q_1, q_2, \ldots, q_n\}$.

We will associate every R^n with the following orthonormal basis:

$$i_1 = \begin{bmatrix} 1 \\ 0 \\ 0 \\ \vdots \\ 0 \\ 0 \end{bmatrix} , \quad i_2 = \begin{bmatrix} 0 \\ 1 \\ 0 \\ \vdots \\ 0 \\ 0 \end{bmatrix} , \quad \cdots , \quad i_{n-1} = \begin{bmatrix} 0 \\ 0 \\ 0 \\ \vdots \\ 1 \\ 0 \end{bmatrix} , \quad i_n = \begin{bmatrix} 0 \\ 0 \\ 0 \\ \vdots \\ 0 \\ 1 \end{bmatrix} . \tag{2.8}$$

With respect to this basis, we have:

$$x = \begin{bmatrix} x_1 \\ x_2 \\ \vdots \\ x_n \end{bmatrix} = x_1 i_1 + x_2 i_2 + \cdots + x_n i_n = I_n \begin{bmatrix} x_1 \\ x_2 \\ \vdots \\ x_n \end{bmatrix},$$

where I_n is the $n \times n$ unit matrix. In other words, the representation of any vector x with respect to the orthonormal basis in (2.8) equals itself.

Norms of Vectors

The concept of norm is a generalization of length or magnitude. Any real valued function of x, denoted by $\|x\|$, can be defined as a norm if it has the following properties:
(i) $\|x\| \geq 0$ for every x and $\|x\| = 0$ if, and only if, $x = 0$
(ii) $\|ax\| = |a| \|x\|$, for any real a
(iii) $\|x_1 + x_2\| \leq \|x_1\| + \|x_2\|$ for every x_1 and x_2

Orthonormalization

A vector x is said to be normalized if its Eucildeam norm is 1 or $x^T x = 1$. Note that $x^T x$ is scalar and xx^T is $n \times n$. Two vectors, x_1 and x_2, are said to be orthogonal if $x_1^T x_2 = x_2^T x_1 = 0$. A set of vectors x_i, $i = 1, 2, \ldots, m$, are said to be orthonormal if:

$$x_i^T x_j = \begin{cases} 0 & \text{if } i \neq j \\ 1 & \text{if } i = j. \end{cases}$$

Given a set of linearly independent vectors $e_1, e_2, \ldots, e_m$, we can obtain an orthonormal set using the procedure that follows:

$$u_1 = e_1 \qquad\qquad\qquad q_1 = u_1 / \|u_1\|$$
$$u_2 = e_2 - (q_1^T e_2) q_1 \qquad\qquad q_2 = u_2 / \|u_2\|$$

$$\vdots \qquad\qquad\qquad\qquad \vdots$$

$$u_m = e_m - \sum_{k=1}^{m-1} (q_k^T e_m) q_k \qquad q_m = u_m / \|u_m\|.$$

The first equation normalizes the vector e_1 to have norm 1. The vector $(q_1^T e_2) q_1$ is the projection of the vector e_2 along q_1. Its subtraction from e_2 yields the vertical part u_2. It is then normalized to 1. Using this procedure, we can obtain an orthonormal set. This is called the *Schmidt orthonormalization procedure*.

Let $A = [a_1\ a_2\ \ldots\ a_m]$ be an $n \times m$ matrix with $m \leq n$. If all columns of A or $\{a_i, i = 1, 2, \ldots, m\}$ are orthonormal, then:

$$A^\mathrm{T} A = \begin{bmatrix} a_1^\mathrm{T} \\ a_2^\mathrm{T} \\ \vdots \\ a_m^\mathrm{T} \end{bmatrix} \begin{bmatrix} a_1 & a_2 & \ldots & a_m \end{bmatrix} = \begin{bmatrix} 1 & 0 & \ldots & 0 \\ 0 & 1 & \ldots & 0 \\ \vdots & \vdots & \ddots & \vdots \\ 0 & 0 & \ldots & 1 \end{bmatrix} = I_m ,$$

where I_m is the unit matrix of order m. Note that, in general, $A A^\mathrm{T} \neq I_n$.

Similarity Transformation

Consider a $n \times n$ matrix A. It maps R^n into itself. If we associate R^n with the orthonormal basis $\{i_1, i_2, \ldots, i_n\}$ in (2.8), then the ith column of A is the representation of $A i_i$, with respect to the orthonormal basis. Now, if we select a different set of basis $\{q_1, q_2, \ldots, q_n\}$, then the matrix A has a different representation $\overline{A}$. It turns out that the ith column of $\overline{A}$ is the representation of $A q_i$, with respect to the basis $\{q_1, q_2, \ldots, q_n\}$. This is illustrated by the following example.

Example 2.1. Consider the matrix:

$$A = \begin{bmatrix} 3 & 2 & -1 \\ -2 & 1 & 0 \\ 4 & 3 & 1 \end{bmatrix} . \tag{2.9}$$

If $b = [0\ 0\ 1]^\mathrm{T}$, we have:

$$Ab = \begin{bmatrix} -1 \\ 0 \\ 1 \end{bmatrix} , \quad A^2 b = A(Ab) = \begin{bmatrix} -4 \\ 2 \\ -3 \end{bmatrix} , \quad A^3 b = A(A^2 b) = \begin{bmatrix} -5 \\ 10 \\ -13 \end{bmatrix} .$$

It can be verified that the following relation holds:

$$A^3 b = 17 b - 15 Ab + 5 A^2 b . \tag{2.10}$$

Because the three vectors b, Ab, and $A^2 b$ are linearly independent, they can be used as a basis. We now compute the representation of A with respect to this basis. It is clear that:

$$A(b) = \begin{bmatrix} b & Ab & A^2 b \end{bmatrix} \begin{bmatrix} 0 \\ 1 \\ 0 \end{bmatrix}$$

$$A(Ab) = \begin{bmatrix} b & Ab & A^2 b \end{bmatrix} \begin{bmatrix} 0 \\ 0 \\ 1 \end{bmatrix}$$

$$A(A^2 b) = \begin{bmatrix} b & Ab & A^2 b \end{bmatrix} \begin{bmatrix} 17 \\ -15 \\ 5 \end{bmatrix} ,$$

where the last equation is obtained from (2.10). Thus, the representation of A, with respect to the basis $\{b, Ab, A^2b\}$, is:

$$\overline{A} = \begin{bmatrix} 0 & 0 & 17 \\ 1 & 0 & -15 \\ 0 & 1 & 5 \end{bmatrix}. \tag{2.11}$$

The preceding discussion can be extended to general cases. For example, suppose that A is a $n \times n$ matrix. If there exists a $n \times 1$ vector b such that the n vectors $b, Ab, \ldots, A^{n-1}b$ are linearly independent, and if

$$A^n = \beta_1 b + \beta_2 Ab + \cdots + \beta_n A^{n-1}b,$$

the representation of A with respect to the basis $\{b, Ab, \ldots, A^{n-1}b\}$ is:

$$\overline{A} = \begin{bmatrix} 0 & 0 & \cdots & 0 & \beta_1 \\ 1 & 0 & \cdots & 0 & \beta_2 \\ 0 & 1 & \cdots & 0 & \beta_3 \\ \vdots & \vdots & & \vdots & \vdots \\ 0 & 0 & \cdots & 0 & \beta_{n-1} \\ 0 & 0 & \cdots & 1 & \beta_n \end{bmatrix}. \tag{2.12}$$

This matrix is said to be in a companion form.

Consider the equation:

$$Ax = y. \tag{2.13}$$

The square matrix A maps x in R^n into y in R^n. With respect to the basis $\{q_1, q_2, \ldots, q_n\}$, the equation becomes:

$$\overline{A}\overline{x} = \overline{y}, \tag{2.14}$$

where $\overline{x}$ and $\overline{y}$ are the representations of x and y, with respect to the basis $\{q_1, q_2, \ldots, q_n\}$. As discussed in (2.7), they are related by:

$$x = Q\overline{x}, \quad y = Q\overline{y}$$

with

$$Q = \begin{bmatrix} q_1 & q_2 & \cdots & q_n \end{bmatrix} \tag{2.15}$$

to be a $n \times n$ nonsingular matrix. Substituting these into (2.13) yields:

$$AQ\overline{x} = Q\overline{y} \quad \text{or} \quad Q^{-1}AQ\overline{x} = \overline{y}. \tag{2.16}$$

Comparing this with (2.14) yields:

$$\overline{A} = Q^{-1}AQ \quad \text{or} \quad A = Q\overline{A}Q^{-1}. \tag{2.17}$$

This is called the similarity transformation and A and $\overline{A}$ are said to be similar. We write (2.17) as:

$$AQ = Q\overline{A} ,$$

or as:

$$A \begin{bmatrix} q_1 & q_2 & \cdots & q_n \end{bmatrix} = \begin{bmatrix} Aq_1 & Aq_2 & \cdots & Aq_n \end{bmatrix} = \begin{bmatrix} q_1 & q_2 & \cdots & q_n \end{bmatrix} \overline{A} .$$

This shows that the ith column of $\overline{A}$ is indeed the representation of Aq_i, with respect to the basis $\{q_1, q_2, \ldots, q_n\}$.

2.2 Transform to Diagonal Form and Jordan Form

A square matrix A has different representations with respect to different sets of basis. In this section, we introduce a set of basis so that the representation will be diagonal or block diagonal.

A real or complex number λ is called an eigenvalue of the $n \times n$ real matrix A if there exists a nonzero vector x, such that $Ax = \lambda x$ is called a (right) eigenvector of A associated with eigenvalue λ. In order to find the eigenvalue of A, we write $Ax = \lambda x = \lambda I x$ as:

$$(A - \lambda I) x = 0 , \tag{2.18}$$

where I is the unit matrix of order n. This is a homogeneous equation. If the matrix $(A - \lambda I)$ is nonsingular, then the only solution of (2.18) is $x = 0$. Thus, in order for (2.18) to have a nonzero solution x, the matrix $(A - \lambda I)$ must be singular or have a determinant. We define:

$$\Delta(\lambda) = \det (\lambda I - A) .$$

It is a monic polynomial of degree n with real coefficients and is called the characteristic polynomical of A. A polynomical is called monic if its leading coefficient is 1. If λ is a root of the characteristic polynomical, then the determinant of $(A - \lambda I)$ is 0 and (2.18) has at least one nonzero solution. Thus, every root of $\Delta(\lambda)$ is an eigenvalue of A. Because $\Delta(\lambda)$ has degree n, the $n \times n$ matrix A has n eigenvalues (not necessarily all distinct).

We mention that the matrices

$$\begin{bmatrix} 0 & 0 & 0 & -\alpha_4 \\ 1 & 0 & 0 & -\alpha_3 \\ 0 & 1 & 0 & -\alpha_2 \\ 0 & 0 & 1 & -\alpha_1 \end{bmatrix} \qquad \begin{bmatrix} -\alpha_1 & -\alpha_2 & -\alpha_3 & -\alpha_4 \\ 1 & 0 & 0 & 0 \\ 0 & 1 & 0 & 0 \\ 0 & 0 & 1 & 0 \end{bmatrix}$$

and their transposes

$$\begin{bmatrix} 0 & 1 & 0 & 0 \\ 0 & 0 & 1 & 0 \\ 0 & 0 & 0 & 1 \\ -\alpha_4 & -\alpha_3 & -\alpha_2 & -\alpha_1 \end{bmatrix} \qquad \begin{bmatrix} -\alpha_1 & 1 & 0 & 0 \\ -\alpha_2 & 0 & 1 & 0 \\ -\alpha_3 & 0 & 0 & 1 \\ -\alpha_4 & 0 & 0 & 0 \end{bmatrix}$$

have the following characteristic polynomial:

$$\Delta(\lambda) = \lambda^4 + \alpha_1\lambda^3 + \alpha_2\lambda^2 + \alpha_3\lambda + \alpha_4 .$$

These matrices can easily be formed from the coefficients of $\Delta(\lambda)$ and are called companion form matrices. The companion form matrices will arise repeatedly later. The matrix in (2.12) is such a form.

Eigenvalues of A Are All Distinct

Suppose λ_i, $i = 1, 2, \ldots, n$, are the eigenvalues of A and that all are distinct. Let q_i be an eigenvector of A associated with λ_i; that is, $Aq_i = \lambda_i q_i$. The set of eigenvectors $\{q_1, q_2, \ldots, q_n\}$ would then be linearly independent and can be used as a basis. Let $\widehat{A}$ be the representation of A with respect to this basis. Then the first column of $\widehat{A}$ is the representation of $Aq_1 = \lambda_1 q_1$ with respect to $\{q_1, q_2, \ldots, q_n\}$. From

$$Aq_1 = \lambda_1 q_1 = \begin{bmatrix} q_1 & q_2 & \cdots & q_n \end{bmatrix} \begin{bmatrix} \lambda_1 \\ 0 \\ 0 \\ \vdots \\ 0 \end{bmatrix},$$

we conclude that the first column of $\widehat{A}$ is $[\lambda_1 \ 0 \ \cdots \ 0]^T$. The second column of $\widehat{A}$ is the representation of $Aq_2 = \lambda_2 q_2$ with respect to $\{q_1, q_2, \ldots, q_n\}$. That is, $[0 \ \lambda_2 \ 0 \ \cdots \ 0]^T$. Proceeding forward, we can establish:

$$\widehat{A} = \begin{bmatrix} \lambda_1 & 0 & 0 & \cdots & 0 \\ 0 & \lambda_2 & 0 & \cdots & 0 \\ 0 & 0 & \lambda_3 & \cdots & 0 \\ \vdots & \vdots & \vdots & & \vdots \\ 0 & 0 & 0 & \cdots & \lambda_N \end{bmatrix}. \tag{2.19}$$

This is a diagonal matrix. Thus, we conclude that every matrix with distinct eigenvalues has a diagonal matrix representation by using its eigenvectors as a basis. Different orderings of eigenvectors will yield different diagonal matrices for the same A.

If we define:

$$Q = \begin{bmatrix} q_1 & q_2 & \cdots & q_n \end{bmatrix},$$ (2.20)

then the matrix $\widehat{A}$ equals:

$$\widehat{A} = Q^{-1}AQ,$$ (2.21)

as derived in (2.17). Computing (2.21) by hand is not simple because of the need to compute the inverse of Q. However, if we know $\widehat{A}$, then we can verify (2.20) by checking $Q\widehat{A} = AQ$.

Example 2.2. Consider the matrix:

$$A = \begin{bmatrix} 0 & 0 & 0 \\ 1 & 0 & 2 \\ 0 & 1 & 1 \end{bmatrix}.$$

Its characteristic polynomial is:

$$\Delta(\lambda) = \det(\lambda I - A) = \det \begin{bmatrix} \lambda & 0 & 0 \\ -1 & \lambda & -2 \\ 0 & -1 & \lambda - 1 \end{bmatrix}$$

$$= \lambda[\lambda(\lambda - 1) - 2] = (\lambda - 2)(\lambda + 1)\lambda.$$

Thus, the eigenvalues of A are 2, −1, and 0. The eigenvector associated with $\lambda = 2$ is any nonzero solution of

$$(A - 2I)q_1 = \begin{bmatrix} -2 & 0 & 0 \\ 1 & -2 & 2 \\ 0 & 1 & -1 \end{bmatrix} q_1 = 0.$$

As such, $q_1 = [0\ 1\ 1]^T$ is an eigenvector associated with $\lambda = 2$. Note that the eigenvector is not unique; $[0\ \alpha\ \alpha]^T$ for any nonzero real α can also be chosen as an eigenvector. The eigenvector associated with $\lambda = -1$ is any nonzero solution of

$$(A - (-1)I)q_2 = \begin{bmatrix} 1 & 0 & 0 \\ 1 & 1 & 2 \\ 0 & 1 & 2 \end{bmatrix} q_2 = 0,$$

which yields $q_2 = [0\ -2\ 1]^T$. Similarly, the eigenvector associated with $\lambda = 0$ can be computed as $q_3 = [2\ 1\ -1]^T$. Therefore, the representation of A, with respect to $\{q_1, q_2, q_3\}$, is:

$$\widehat{A} = \begin{bmatrix} 2 & 0 & 0 \\ 0 & -1 & 0 \\ 0 & 0 & 0 \end{bmatrix}.$$ (2.22)

It is a diagonal matrix with eigenvalues on the diagonal. This matrix can also be obtained by computing

$$\widehat{A} = Q^{-1}AQ$$

with

$$Q = \begin{bmatrix} q_1 & q_2 & q_3 \end{bmatrix} = \begin{bmatrix} 0 & 0 & 2 \\ 1 & -2 & 1 \\ 1 & 1 & -1 \end{bmatrix}.$$

However, it is simpler to verify $Q\widehat{A} = AQ$ or:

$$\begin{bmatrix} 0 & 0 & 2 \\ 1 & -2 & 1 \\ 1 & 1 & -1 \end{bmatrix} \begin{bmatrix} 2 & 0 & 0 \\ 0 & -1 & 0 \\ 0 & 0 & 0 \end{bmatrix} = \begin{bmatrix} 0 & 0 & 0 \\ 1 & 0 & 2 \\ 0 & 1 & 1 \end{bmatrix} \begin{bmatrix} 0 & 0 & 2 \\ 1 & -2 & 1 \\ 1 & 1 & -1 \end{bmatrix}.$$

Eigenvalues of *A* Are Not All Distinct

An eigenvalue with multiplicity 2 or higher is called a repeated eigenvalue. In contrast, an eigenvalue with multiplicity 1 is called a simple eigenvalue. If A has only simple eigenvalues, it always has a diagonal form representation. If A has repeated eigenvalues, then it may not have a diagonal form representation. However, it has a block diagonal and triangular form representation, as we will discuss next.

Consider a $n \times n$ matrix A with eigenvalue λ and multiplicity n. In other words, A has only one distinct eigenvalue. To simplify the discussion, we assume $n = 4$. Suppose the matrix $(A - \lambda I)$ has rank $n - 1 = 3$ or, equivalently, nullity 1, then the equation

$$(A - \lambda I)\, q = 0$$

has only one independent solution. Thus, A has only one eigenvector associated with λ. We need $n - 1 = 3$ more linearly independent vectors to form a basis for R^4. The three vectors q_2, q_3, q_4 will be chosen to have the properties $(A - \lambda I)^2 q_2 = 0$, $(A - \lambda I)^3 q_3 = 0$, and $(A - \lambda I)^4 q_4 = 0$.

A vector v is called a *generalized eigenvector* of grade n if

$$(A - \lambda I)^n\, v = 0$$

and

$$(A - \lambda I)^{n-1}\, v \neq 0 .$$

If $n = 1$, they reduce to $(A - \lambda I)v = 0$ and $v \neq 0$ (v is an ordinary eigenvector). For $n = 4$, we define:

$$v_4 = v$$
$$v_3 = (A - \lambda I)\, v_4 = (A - \lambda I)\, v$$
$$v_2 = (A - \lambda I)\, v_3 = (A - \lambda I)^2\, v$$
$$v_1 = (A - \lambda I)\, v_2 = (A - \lambda I)^3\, v .$$

They are called a chain of generalized eigenvectors of length $n = 4$ and have the properties $(A - \lambda I)v_1 = 0$, $(A - \lambda I)^2 v_2 = 0$, $(A - \lambda I)^3 v_3 = 0$, and $(A - \lambda I)^4 v_4 = 0$. These vectors, as generated, are automatically linearly independent and can be used as a basis. From these equations, we can readily obtain:

$$Av_1 = \lambda v_1$$
$$Av_2 = v_1 + \lambda v_2$$
$$Av_3 = v_2 + \lambda v_3$$
$$Av_4 = v_3 + \lambda v_4 .$$

Then the representation of A with respect to the basis $\{v_1, v_2, v_3, v_4\}$ is:

$$J = \begin{bmatrix} \lambda & 1 & 0 & 0 \\ 0 & \lambda & 1 & 0 \\ 0 & 0 & \lambda & 1 \\ 0 & 0 & 0 & \lambda \end{bmatrix}. \tag{2.23}$$

We verify this for the first and last columns. The first column of J is the representation of $Av_1 = \lambda v_1$, with respect to $\{v_1, v_2, v_3, v_4\}$, which is $[\lambda\ 0\ 0\ 0]^T$. The last column of J is the representation of $Av_4 = v_3 + \lambda v_4$ with respect to $\{v_1, v_2, v_3, v_4\}$, which is $[0\ 0\ 1\ \lambda]^T$. This verifies the representation in (2.23). The matrix J has eigenvalues on the diagonal and 1 on the superdiagonal. If we reverse the order of the basis, then the 1 in (2.23) will appear on the subdiagonal. The matrix is called a Jordan block of order $n = 4$.

If $(A - \lambda I)$ has rank $n - 2$ or, equivalently, nullity 2, then the equation

$$(A - \lambda I)\,q = 0$$

has two linearly independent solutions. Thus, A has two linearly independent eigenvectors and we need $(n - 2)$ generalized eigenvectors. In this case, two chains of generalized eigenvectors exist; $\{v_1, v_2, \ldots, v_k\}$ and $\{u_1, u_2, \ldots, u_l\}$ with $k + l = n$. If v_1 and u_2 are linearly independent, then the set of n vectors $\{v_1, \ldots, v_k, u_1, \ldots, u_l\}$ is linearly independent and can be used as a basis. With respect to this basis, the representation of A is a block diagonal matrix of form:

$$\widehat{A} = \mathrm{diag}\left\{J_1,\ \ J_2\right\},$$

where J_1 and J_2 are, respectively, Jordan blocks of order k and l.

Now we discuss a specific example. Consider a 5×5 matrix A with repeated eigenvalue λ_1 with multiplicity 4 and simple eigenvalue λ_2. Suppose that a nonsingular matrix Q exists, such that

$$\widehat{A} = Q^{-1}AQ$$

assumes one of the following forms:

$$\widehat{A}_1 = \begin{bmatrix} \lambda_1 & 1 & 0 & 0 & 0 \\ 0 & \lambda_1 & 1 & 0 & 0 \\ 0 & 0 & \lambda_1 & 1 & 0 \\ 0 & 0 & 0 & \lambda_1 & 0 \\ 0 & 0 & 0 & 0 & \lambda_2 \end{bmatrix}, \quad \widehat{A}_2 = \begin{bmatrix} \lambda_1 & 1 & 0 & 0 & 0 \\ 0 & \lambda_1 & 1 & 0 & 0 \\ 0 & 0 & \lambda_1 & 0 & 0 \\ 0 & 0 & 0 & \lambda_1 & 0 \\ 0 & 0 & 0 & 0 & \lambda_2 \end{bmatrix},$$

$$\widehat{A}_3 = \begin{bmatrix} \lambda_1 & 1 & 0 & 0 & 0 \\ 0 & \lambda_1 & 0 & 0 & 0 \\ 0 & 0 & \lambda_1 & 1 & 0 \\ 0 & 0 & 0 & \lambda_1 & 0 \\ 0 & 0 & 0 & 0 & \lambda_2 \end{bmatrix}, \quad \widehat{A}_4 = \begin{bmatrix} \lambda_1 & 1 & 0 & 0 & 0 \\ 0 & \lambda_1 & 0 & 0 & 0 \\ 0 & 0 & \lambda_1 & 0 & 0 \\ 0 & 0 & 0 & \lambda_1 & 0 \\ 0 & 0 & 0 & 0 & \lambda_2 \end{bmatrix}, \quad (2.24)$$

$$\widehat{A}_5 = \begin{bmatrix} \lambda_1 & 0 & 0 & 0 & 0 \\ 0 & \lambda_1 & 0 & 0 & 0 \\ 0 & 0 & \lambda_1 & 0 & 0 \\ 0 & 0 & 0 & \lambda_1 & 0 \\ 0 & 0 & 0 & 0 & \lambda_2 \end{bmatrix}.$$

The first matrix occurs when the nullity of $(A-\lambda_1 I)$ is 1. If the nullity is 2, then $\widehat{A}$ has two Jordan blocks associated with λ_1; it may assume the form in $\widehat{A}_2$ or $\widehat{A}_3$. If $(A - \lambda_1 I)$ has nullity 3, then $\widehat{A}$ has three Jordan blocks associated with λ_1, as shown in $\widehat{A}_4$. Certainly, the positions of the Jordan blocks can be changed by changing the order of the basis. If the nullity is 4, then $\widehat{A}$ is a diagonal matrix as shown in $\widehat{A}_5$. All these matrices are triangular and block diagonal with Jordan blocks on the diagonal; they are said to be in Jordan form. A diagonal matrix is a degenerated Jordan form; its Jordan blocks all have order 1.

Jordan form matrices are triangular and block diagonal and can be used to establish many general properties of matrices. For example, because $\det(CD) = \det C \det D$ and $\det Q \det Q^{-1} = \det I = 1$, from $A = Q\widehat{A}Q^{-1}$, we have:

$$\det A = \det Q \det \widehat{A} \det Q^{-1} = \det \widehat{A}.$$

The determinant of $\widehat{A}$ is the product of all diagonal entries or, equivalently, all eigenvalues of A. Thus, we have:

$$\det A = product\ of\ all\ \text{eigenvalues of } A,$$

which implies that A is nonsingular if, and only if, it has nonzero eigenvalue.

We discuss a useful property of Jordan blocks to conclude this section. Consider the Jordan block in (2.23) with order 4. Then we have:

$$(J - \lambda I) = \begin{bmatrix} 0 & 1 & 0 & 0 \\ 0 & 0 & 1 & 0 \\ 0 & 0 & 0 & 1 \\ 0 & 0 & 0 & 0 \end{bmatrix} \qquad (J - \lambda I)^2 = \begin{bmatrix} 0 & 0 & 1 & 0 \\ 0 & 0 & 0 & 1 \\ 0 & 0 & 0 & 0 \\ 0 & 0 & 0 & 0 \end{bmatrix}$$

$$(J - \lambda I)^3 = \begin{bmatrix} 0 & 0 & 0 & 1 \\ 0 & 0 & 0 & 0 \\ 0 & 0 & 0 & 0 \\ 0 & 0 & 0 & 0 \end{bmatrix} ,$$

(2.25)

and $(J - \lambda I)^k = 0$ – for $k \geq 4$. This is called nilpotent.

Example 2.3. Consider the matrix:

$$A = \begin{bmatrix} 0 & 1 & -1 \\ -6 & -11 & 6 \\ -6 & -11 & 5 \end{bmatrix} .$$

Its characteristic polynomial is:

$$\Delta(\lambda) = \det(\lambda I - A) = \det \begin{bmatrix} \lambda & -1 & 1 \\ 6 & \lambda + 11 & -6 \\ 6 & 11 & \lambda - 5 \end{bmatrix} = (\lambda + 1)(\lambda + 2)(\lambda + 3) .$$

Thus, the eigenvalues of A are -1, -2, and -3.

$$q_1 = \begin{bmatrix} 1 & 0 & 1 \end{bmatrix}^T ,$$
$$q_2 = \begin{bmatrix} 1 & 2 & 4 \end{bmatrix}^T ,$$
$$q_3 = \begin{bmatrix} 1 & 6 & 9 \end{bmatrix}^T .$$

We can obtain the diagonal matrix by computing:

$$\widehat{A} = Q^{-1} A Q ,$$

with:

$$Q = \begin{bmatrix} q_1 & q_2 & q_3 \end{bmatrix} = \begin{bmatrix} 1 & 1 & 1 \\ 0 & 2 & 6 \\ 1 & 4 & 9 \end{bmatrix} .$$

Then we can get:

$$Q^{-1} = \frac{1}{9} \begin{bmatrix} 2 & -5 & 2 \\ 6 & 3 & -3 \\ 1 & 2 & 1 \end{bmatrix} .$$

We have the diagonal form:

$$\widehat{A} = \begin{bmatrix} -1 & 0 & 0 \\ 0 & -2 & 0 \\ 0 & 0 & -3 \end{bmatrix},$$

$$\widehat{B} = Q^{-1}B = \begin{bmatrix} -2 \\ 3 \\ -1 \end{bmatrix},$$

$$\widehat{C} = CQ = \begin{bmatrix} 1 & 1 & 1 \end{bmatrix}.$$

The result in this example can easily be obtained using MATLAB. Typing:

```
a=[0 1 -1;-6 -11 6;-6 -11 5];
[q,d]=eig(a)
```

yields

```
q =
    0.7071    -0.2182    -0.0921
    0.0000    -0.4364    -0.5523
    0.7071    -0.8729    -0.8285
d =
   -1.0000      0          0
    0         -2.0000      0
    0          0         -3.0000,
```

where d is the diagonal matrix. The matrix is different from the Q, but their corresponding columns differ only by a constant. This is due to the nonuniqueness of eigenvectors and how every column of q is normalized to have norm 1 in MATLAB.

Example 2.4. Try to transform the state space representation to the Jordan block:

$$\dot{x} = \begin{pmatrix} 0 & 1 & 0 \\ 0 & 0 & 1 \\ 2 & 3 & 0 \end{pmatrix} x + \begin{pmatrix} 0 \\ 0 \\ 1 \end{pmatrix} u .$$

$$y = \begin{pmatrix} 1 & 0 & 0 \end{pmatrix} x$$

Solution. First, we get the eigenvalues of A:

$$|\lambda I - A| = \begin{vmatrix} \lambda & -1 & 0 \\ 0 & \lambda & -1 \\ -2 & -3 & \lambda \end{vmatrix} = 0 .$$

That is,

$$\lambda^3 - 3\lambda - 2 = 0 .$$

We get:

$$\lambda_{1,2} = -1 , \quad \lambda_3 = 2 .$$

Eigenvector Q_1 corresponding to $\lambda_1 = -1$:

$$\begin{pmatrix} 0 & 1 & 0 \\ 0 & 0 & 1 \\ 2 & 3 & 0 \end{pmatrix} \begin{pmatrix} q_{11} \\ q_{21} \\ q_{31} \end{pmatrix} = - \begin{pmatrix} q_{11} \\ q_{21} \\ q_{31} \end{pmatrix}.$$

Then:

$$Q_1 = \begin{pmatrix} q_{11} \\ q_{21} \\ q_{31} \end{pmatrix} = \begin{pmatrix} 1 \\ -1 \\ 1 \end{pmatrix}.$$

Eigenvector Q_2 corresponding to $\lambda_2 = -1$:

$$\lambda_1 Q_2 - A Q_2 = -Q_1,$$

$$-\begin{pmatrix} q_{21} \\ q_{22} \\ q_{23} \end{pmatrix} - \begin{pmatrix} 0 & 1 & 0 \\ 0 & 0 & 1 \\ 2 & 3 & 0 \end{pmatrix} \begin{pmatrix} q_{21} \\ q_{22} \\ q_{23} \end{pmatrix} = - \begin{pmatrix} 1 \\ -1 \\ 1 \end{pmatrix},$$

$$Q_2 = \begin{pmatrix} 1 \\ 0 \\ -1 \end{pmatrix}.$$

Last, eigenvector Q_3 corresponding to $\lambda_3 = 2$:

$$\lambda_3 Q_3 = A Q_3,$$

$$Q_3 = \begin{pmatrix} 1 \\ 2 \\ 4 \end{pmatrix},$$

$$T = \begin{pmatrix} Q_1 & Q_2 & Q_3 \end{pmatrix} = \begin{pmatrix} 1 & 1 & 1 \\ -1 & 0 & 2 \\ 1 & -1 & 4 \end{pmatrix}.$$

It can be calculated as:

$$T^{-1} = \frac{1}{9} \begin{pmatrix} 2 & -5 & 2 \\ 6 & 3 & -3 \\ 1 & 2 & 1 \end{pmatrix}.$$

Then, we can calculate the matrix we want:

$$J = \begin{pmatrix} -1 & 1 & 0 \\ 0 & -1 & 0 \\ 0 & 0 & 2 \end{pmatrix},$$

$$T^{-1} B = \begin{pmatrix} \frac{2}{9} \\ -\frac{1}{3} \\ \frac{1}{9} \end{pmatrix},$$

$$CT = \begin{pmatrix} 1 & 1 & 1 \end{pmatrix}.$$

Using MATLAB, typing:

```
A=[0 1 0;0 0 1;2 3 0];
[v,J] = jordan(A);
```

yields

```
v =
     0.1111      0.6667      0.8889
     0.2222     -0.6667     -0.2222
     0.4444      0.6667     -0.4444
J =
     2     0     0
     0    -1     1
     0     0    -1
```

jordan(A) computes the Jordan Canonical/Normal Form of the matrix A. The columns of V are the generalized eigenvectors. J is the Jordan canonical form.

Exercise

2.1. Judge whether the following vectors are linearly dependent or not.

(1) $\begin{pmatrix} -1 \\ 3 \\ 1 \end{pmatrix}, \begin{pmatrix} 2 \\ 1 \\ 0 \end{pmatrix}, \begin{pmatrix} 1 \\ 4 \\ 1 \end{pmatrix}.$ (2) $\begin{pmatrix} 2 \\ 3 \\ 0 \end{pmatrix}, \begin{pmatrix} -1 \\ 4 \\ 0 \end{pmatrix}, \begin{pmatrix} 0 \\ 0 \\ 2 \end{pmatrix}.$

2.2. Find the characteristic polynomials of the following matrices.

(1) $\begin{pmatrix} \lambda_1 & 1 & 0 & 0 \\ 0 & \lambda_1 & 1 & 0 \\ 0 & 0 & \lambda_1 & 0 \\ 0 & 0 & 0 & \lambda_2 \end{pmatrix}.$ (2) $\begin{pmatrix} \lambda_1 & 1 & 0 & 0 \\ 0 & \lambda_1 & 1 & 0 \\ 0 & 0 & \lambda_1 & 0 \\ 0 & 0 & 0 & \lambda_1 \end{pmatrix}.$

(3) $\begin{pmatrix} \lambda_1 & 1 & 0 & 0 \\ 0 & \lambda_1 & 0 & 0 \\ 0 & 0 & \lambda_1 & 0 \\ 0 & 0 & 0 & \lambda_1 \end{pmatrix}.$ (4) $\begin{pmatrix} \lambda_1 & 0 & 0 & 0 \\ 0 & \lambda_1 & 0 & 0 \\ 0 & 0 & \lambda_1 & 0 \\ 0 & 0 & 0 & \lambda_1 \end{pmatrix}.$

2.3. Find eigenvectors of the following matrices.

(1) $A = \begin{pmatrix} -2 & 1 \\ -1 & -2 \end{pmatrix}.$ (2) $A = \begin{pmatrix} 0 & 1 \\ -6 & -5 \end{pmatrix}.$

(3) $A = \begin{pmatrix} 0 & 1 & 0 \\ 3 & 0 & 2 \\ -12 & -7 & -6 \end{pmatrix}.$ (4) $A = \begin{pmatrix} 1 & 2 & -1 \\ -1 & 0 & -1 \\ 4 & 4 & 5 \end{pmatrix}.$

2.4. find jordan form representations of the following matrices:

$$A_1 = \begin{bmatrix} 1 & 4 & 10 \\ 0 & 2 & 0 \\ 0 & 0 & 3 \end{bmatrix}, \quad A_2 = \begin{bmatrix} 0 & 1 & 0 \\ 0 & 0 & 1 \\ -2 & -4 & -3 \end{bmatrix},$$

$$A_3 = \begin{bmatrix} 1 & 0 & -1 \\ 0 & 1 & 0 \\ 0 & 0 & 2 \end{bmatrix}, \quad A_4 = \begin{bmatrix} 0 & 4 & 3 \\ 0 & 20 & 16 \\ 0 & -25 & -20 \end{bmatrix}.$$

2.5. Find Jordan form representations of the following state space equations:

(1) $$\begin{pmatrix} \dot{x}_1 \\ \dot{x}_2 \end{pmatrix} = \begin{pmatrix} -2 & 1 \\ 1 & -2 \end{pmatrix} \begin{pmatrix} x_1 \\ x_2 \end{pmatrix} + \begin{pmatrix} 0 \\ 1 \end{pmatrix} u$$

$$y = \begin{pmatrix} 1 & 0 \end{pmatrix} \begin{pmatrix} x_1 \\ x_2 \end{pmatrix}.$$

(2) $$\begin{pmatrix} \dot{x}_1 \\ \dot{x}_2 \\ \dot{x}_3 \end{pmatrix} = \begin{pmatrix} 4 & 1 & -2 \\ 1 & 0 & 2 \\ 1 & -1 & 3 \end{pmatrix} \begin{pmatrix} x_1 \\ x_2 \\ x_3 \end{pmatrix} + \begin{pmatrix} 3 & 1 \\ 2 & 7 \\ 5 & 3 \end{pmatrix} u$$

$$y = \begin{pmatrix} 1 & 2 & 0 \\ 0 & 1 & 1 \end{pmatrix} \begin{pmatrix} x_1 \\ x_2 \\ x_3 \end{pmatrix}.$$

3 Solution of State Space Model

3.1 Introduction

In Chapter 2, it was shown that linear systems can be described by state space equations. This chapter will discuss the solution of the state space equation for LTI systems. Different methods to solve the state transition matrix for both continuous systems and discrete systems are discussed in detail.

3.2 Solution of LTI State Equations

Consider the LTI state space equation:

$$\dot{x}(t) = Ax(t) + Bu(t) , \tag{3.1}$$
$$y(t) = Cx(t) + Du(t) , \tag{3.2}$$

where A, B are, respectively, $n \times n$, $n \times p$ dimensional constant matrices.

The problem is to find the solution excited by the initial state $x(t)|_{t=0} = x(0)$ and the input $u(t)$. The solution hinges on the exponential function of A. In particular, the following property

$$\frac{d}{dt} e^{At} = Ae^{At} = e^{At}A \tag{3.3}$$

is necessary to develop the solution.

Rewrite the equation as:

$$\dot{x}(t) - Ax(t) = Bu(t) . \tag{3.4}$$

Premultiplying e^{-At} to both sides of (3.3) yields:

$$e^{-At}[\dot{x}(t) - Ax(t)] = e^{-At}Bu(t) , \tag{3.5}$$

which implies that

$$\frac{d}{dt}(e^{-At}x(t)) = e^{-At}Bu(t) .$$

Its integration from 0 to t yields:

$$e^{-A\tau}x(\tau)\Big|_{\tau=0}^{t} = \int_{0}^{t} e^{-A\tau}Bu(\tau)d\tau .$$

Thus, we have:

$$e^{-At}x(t) - e^{0}x(0) = \int_{0}^{t} e^{-A\tau}Bu(\tau)d\tau . \tag{3.6}$$

https://doi.org/10.1515/9783110574951-003

Because the inverse of e^{-At} is e^{At} and $e^0 = I$, equation (3.6) implies:

$$x(t) = e^{At}x(0) + \int_0^t e^{A(t-\tau)}Bu(\tau)d\tau, \tag{3.7}$$

or, equivalently:

$$x(t) = \Phi(t)x(0) + \int_0^t \Phi(t-\tau)Bu(\tau)d\tau, \tag{3.8}$$

where $\Phi(t) = e^{At}$ is called the state transition matrix. Equation (3.7), or (3.8), is the solution of (3.1).

To verify that (3.7) is the solution of (3.1), it is necessary to show that (3.7) satisfies (3.1) and the initial condition $x(t) = x(0)$ at $t = 0$. Indeed, at $t = 0$, (3.7) reduces to

$$x(0) = e^{A \cdot 0}x(0) = e^0 x(0) = Ix(0) = x(0).$$

Thus, (3.6) satisfies the initial condition. The equation

$$\frac{\partial}{\partial t}\int_{t_0}^t f(t, \tau)d\tau = \int_{t_0}^t \left(\frac{\partial}{\partial t}f(t, \tau)\right)d\tau + f(t, \tau)\big|_{\tau=t} \tag{3.9}$$

is needed to show that (3.7) satisfies (3.1). Differentiating (3.7) and using (3.9) obtains:

$$\dot{x}(t) = \frac{d}{dt}\left[e^{At}x(0) + \int_0^t e^{A(t-\tau)}Bu(\tau)d\tau\right]$$

$$= Ae^{At}x(0) + \int_0^t Ae^{A(t-\tau)}Bu(\tau)d\tau + e^{A(t-\tau)}Bu(\tau)\big|_{\tau=t}$$

$$= A\left(e^{At}x(0) + \int_0^t e^{A(t-\tau)}Bu(\tau)d\tau\right) + e^{A \cdot 0}Bu(t).$$

Substituting (3.7) into the above equation results in:

$$\dot{x}(t) = Ax(t) + Bu(t).$$

Thus, (3.7) meets (3.1) and the initial condition $x(0)$, and is the solution of (3.1). Substituting (3.7) into (3.2) yields the solution of (3.2) as:

$$y(t) = Ce^{At}x(0) + C\int_0^t e^{A(t-\tau)}Bu(\tau)d\tau + Du(t). \tag{3.10}$$

This solution and (3.7) are computed directly in the time domain. It is also convenient to compute the solutions by using the Laplace transform. Applying the Laplace transform to (3.1) yields:

$$sX(s) - x(0) = AX(s) + BU(s)$$
$$(sI - A)X(s) = x(0) + BU(s) .$$

Premultiplying $(sI - A)^{-1}$ to both sides of the above equation yields:

$$x(s) = (sI - A)^{-1}x(0) + (sI - A)^{-1}BU(s) . \qquad (3.11)$$

Notice that

$$(sI - A)^{-1} = \ell[\Phi(t)]$$
$$U(s) = \ell[u(t)] .$$

From the property of Laplace transform function, the second term of the right side of (3.11) can be expressed as:

$$(sI - A)^{-1}BU(s) = \ell \left[\int_0^t \Phi(t - \tau)Bu(\tau) \right] d\tau . \qquad (3.12)$$

Substituting (3.12) into (3.11) and applying the inverse Laplace transform yields:

$$x(t) = \Phi(t)x(0) + \int_0^t \Phi(t - \tau)Bu(\tau)d\tau \qquad (3.13)$$

Equation (3.13) is just the time domain solution of (3.1).

Example 3.1. Find the solution of the following system excited by the unit step function:

$$\dot{x} = \begin{bmatrix} 0 & 1 \\ -2 & -3 \end{bmatrix} x + \begin{bmatrix} 0 \\ 1 \end{bmatrix} u .$$

Solution.

$$\Phi(t) = e^{At} = (sI - A)^{-1} = \begin{bmatrix} 2e^{-t} - e^{-2t} & e^{-t} - e^{-2t} \\ -2e^{-t} + 2e^{-2t} & -e^{-t} + 2e^{-2t} \end{bmatrix} .$$

Substituting $B = [0\ 1]^T$ and $u(t) = 1(t)$ into equation (3.8) yields:

$$x(t) = \begin{bmatrix} 2e^{-t} - e^{-2t} & e^{-t} - e^{-2t} \\ -2e^{-t} + 2e^{-2t} & -e^{-t} + 2e^{-2t} \end{bmatrix} \begin{bmatrix} x_1(0) \\ x_2(0) \end{bmatrix} + \int_0^t \begin{bmatrix} e^{-(t-\tau)} - e^{-2(t-\tau)} \\ -e^{-(t-\tau)} + 2e^{-2(t-\tau)} \end{bmatrix} d\tau$$

$$= \begin{bmatrix} (2e^{-t} - e^{-2t})x_1(0) + (e^{-t} - e^{-2t})x_2(0) \\ (-2e^{-t} + 2e^{-2t})x_1(0) + (-e^{-t} + 2e^{-2t})x_2(0) \end{bmatrix} + \begin{bmatrix} \frac{1}{2} - e^{-t} + \frac{1}{2}e^{-2t} \\ e^{-t} - e^{-2t} \end{bmatrix}$$

$$= \begin{bmatrix} \frac{1}{2} + [2x_1(0) + x_2(0) - 1] e^{-t} - \left[x_1(0) + x_2(0) - \frac{1}{2}\right] e^{-2t} \\ -[2x_1(0) + x_2(0) - 1] e^{-t} - [2x_1(0) + 2x_2(0) - 1] e^{-2t} \end{bmatrix} .$$

If the initial condition is zero, i.e., $x(0) = 0$, then the response of the system depends only on the excitation of the control action:

$$\begin{bmatrix} x_1(t) \\ x_2(t) \end{bmatrix} = \begin{bmatrix} \frac{1}{2} - e^{-t} + \frac{1}{2}e^{-2t} \\ e^{-t} - e^{-2t} \end{bmatrix} .$$

In the following, the solution of the state space equation (3.8) is presented under three special control signals.

(1) The impulse response: When $u(t) = K\delta(t)$, $x(0_-) = x_0$,

$$x(t) = e^{At}x_0 + e^{At}BK ,$$

where K is a constant vector having the same dimension with $u(t)$.

(2) The step response: When $u(t) = K \times 1(t)$, $x(0_-) = x_0$,

$$x(t) = e^{At}x_0 + A^{-1}(e^{At} - 1)BK .$$

(3) The scope response: When $u(t) = Kt \times 1(t)$, $x(0_-) = x_0$,

$$x(t) = e^{At}x_0 + \left[A^{-2}(e^{At} - 1) - A^{-1}t \right] BK .$$

3.3 State Transfer Matrix

3.3.1 Properties

Consider the LTI state space equation:

$$\dot{x}(t) = Ax(t) + Bu(t) , \qquad (3.14)$$

where $x(t) \in R^n$, $u(t) \in R^r$, $A \in R^{n \times n}$, $B \in R^{n \times r}$.

The solution of the system is:

$$x(t) = e^{At}x_0 \quad \text{or} \quad x(t) = e^{A(t-t_0)}x_0 ,$$

which reflects a vector transition relation from the initial state vector x_0 to the state $x(t)$ of any $t > 0$, or $t > t_0$, where e^{At} is called a transfer matrix. It is not a constant matrix – it is a $n \times n$ time varying function matrix because the elements of the matrix are general functions of t. This means that it makes the state vector change constantly in the state space, so $\Phi(t) = e^{At}$ is also called state transition matrix. $\Phi(t) = e^{At}$ is the transition matrix from $x(0)$ to $x(t)$, while $\Phi(t - t_0) = e^{A(t-t_0)}$ is the transition matrix from $x(t_0)$ to $x(t)$. Therefore, the solution of $\dot{x}(t) = Ax(t) + Bu(t)$ can also be expressed as:

$$x(t) = \Phi(t)x(0) ,$$

or as:

$$x(t) = \Phi(t - t_0)x(t_0) .$$

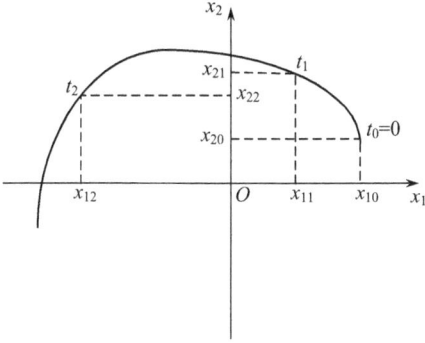

Fig. 3.1: State transition trajectory.

Its geometric meaning, taking two-dimensional state vector for example, can be represented as in Figure 3.1.

From Figure 3.1, we know $x(0) = [x_{10} \ x_{20}]^T$ when $t = 0$. If we consider this as an initial condition and $\Phi(t_1)$ is known, when $t = t_1$, the state will be:

$$x(t_1) = \begin{bmatrix} x_{11} \\ x_{21} \end{bmatrix} = \Phi(t_1)x(0) . \tag{3.15}$$

If $\Phi(t_2)$ is known, when $t = t_2$, the state will be:

$$x(t_2) = \begin{bmatrix} x_{12} \\ x_{22} \end{bmatrix} = \Phi(t_2)x(0) . \tag{3.16}$$

That is to say, the state $x(0)$ will transfer to the state $x(t_1)$ or $x(t_2)$ according to $\Phi(t_1)$ or $\Phi(t_2)$.

If we take $t = t_1$ as initial time, the state $x(t_1)$ is the initial state and the state transited from t_1 to t_2 will be:

$$x(t_2) = \Phi(t_2 - t_1)x(t_1) . \tag{3.17}$$

Substituting $x(t_1)$ of equation (3.15) into the above equation can result in:

$$x(t_2) = \Phi(t_2 - t_1)\Phi(t_1)x(0) . \tag{3.18}$$

Equation (3.18) shows the transformation of the state x from $x(0)$ to $x(t_1)$, and then to $x(t_2)$.

Comparing equation (3.16) and (3.18), it is clear that:

$$\Phi(t_2 - t_1)\Phi(t_1) = \Phi(t_2) ,$$

or that:

$$e^{A(t_2 - t_1)}e^{At_1} = e^{At_2} . \tag{3.19}$$

Such a relation is called the combination property.

From the above, for any given initial state vector, $x(t_0)$ can be transited to $x(t)$ at any t using state transition matrix. In other words, matrix differential equations can be solved in an arbitrary time period. This is another advantage of state space representation on a dynamic system.

Property 1

$$\begin{cases} \Phi(t)\Phi(\tau) = \Phi(t + \tau) \\ \text{or} \quad e^{At}e^{A\tau} = e^{A(t+t)} \end{cases} \tag{3.20}$$

This is a combination property, which means a combination of transition from $-\tau$ to 0 and transition from 0 to t. That is to say,

$$\Phi(t - 0)\Phi[0 - (-\tau)] = \Phi[t - (-\tau)] = \Phi(t + \tau).$$

Property 2

$$\begin{cases} \Phi(t - t) = I \\ e^{A(t-t)} = I \end{cases} \tag{3.21}$$

This property means that when the state vector transit from time instant t to t, the state vector is invariable.

Property 3

$$\begin{cases} [\Phi(t)]^{-1} = \Phi(-t) \\ \text{or} \quad [e^{At}]^{-1} = e^{-At} \end{cases} \tag{3.22}$$

This property shows that the inverse of the transition matrix means the reversion of time. If $x(t)$ is known, we can have $x(t_0)$ at time t while $t_0 < t$.

Property 4
For the transition matrix:

$$\begin{cases} \dot{\Phi}(t) = A\Phi(t) = \Phi(t)A \\ \text{or} \quad \dfrac{d}{dt}e^{At} = Ae^{At} = e^{At} \cdot A \end{cases} \tag{3.23}$$

This property shows $\Phi(t)$ or e^{At} can change with matrix A.

Property 5
For the $n \times n$ square matrices A and B, if and only if $AB = BA$, there exists $e^{At}e^{Bt} = e^{(A+B)t}$; if $AB \neq BA$, then $e^{At}e^{Bt} \neq e^{(A+B)t}$.

This property shows that, unless A and B are matrix exchangeable, the product of their respective matrix exponential functions is not equivalent to matrix exponential function of the sum of A and B.

Below are some special matrix exponential functions.

(1) If A is diagonal matrix,

$$A = \Lambda = \begin{bmatrix} \lambda_1 & & & 0 \\ & \lambda_2 & & \\ & & \ddots & \\ 0 & & & \lambda_n \end{bmatrix},$$

then

$$e^{At} = \Phi(t) = \begin{bmatrix} e^{\lambda_1 t} & & & 0 \\ & e^{\lambda_2 t} & & \\ & & \ddots & \\ 0 & & & e^{\lambda_n t} \end{bmatrix}. \tag{3.24}$$

(2) If A can be diagonalized through nonsingular transformation,

$$T^{-1}AT = \Lambda,$$

then

$$e^{At} = \Phi(t) = T \begin{bmatrix} e^{\lambda_1 t} & & & 0 \\ & e^{\lambda_2 t} & & \\ & & \ddots & \\ 0 & & & e^{\lambda_n t} \end{bmatrix} T^{-1}. \tag{3.25}$$

(3) If A is Jordan matrix,

$$A = J = \begin{bmatrix} \lambda & 1 & & & & 0 \\ & \lambda & 1 & & & \\ & & \lambda & 1 & & \\ & & & \lambda & 1 & \\ & & & & \lambda & 1 \\ 0 & & & & & \lambda \end{bmatrix}, \tag{3.26}$$

then

$$e^{Jt} = \Phi(t) = e^{\lambda t} \begin{bmatrix} 1 & t & \frac{1}{2!}t^2 & \cdots & \frac{1}{(n-1)!}t^{n-1} \\ 0 & 1 & t & \cdots & \frac{1}{(n-2)!}t^{n-2} \\ & & & \ddots & \\ 0 & 0 & 0 & \cdots & t \\ 0 & 0 & 0 & \cdots & 1 \end{bmatrix}. \tag{3.27}$$

(4) If

$$A = \begin{bmatrix} \sigma & \omega \\ -\omega & \sigma \end{bmatrix},$$

then

$$e^{At} = \Phi(t) = \begin{bmatrix} \cos \omega t & \sin \omega t \\ -\sin \omega t & \cos \omega t \end{bmatrix}. \tag{3.28}$$

3.3.2 Calculating the State Transition Matrix

For calculating the matrix e^{At}, many methods have been proposed. Here are the four most popular ones.

Method 1: Expansion of e^{At}

This method takes place entirely in the time domain and is based on the expansion of e^{At} in a power series. Namely, it is based on the definition of e^{At} or $\Phi(t)$:

$$e^{At} = I + At + \frac{A^2 t^2}{2!} + \frac{A^3 t^3}{3!} + \cdots = \sum_{k=0}^{\infty} \frac{1}{k!} A^k t^k . \tag{3.29}$$

Example 3.2. Compute the matrix e^{At}, where $A = \begin{bmatrix} 0 & 1 \\ -2 & -3 \end{bmatrix}$.

Solution.

$$e^{At} = \begin{bmatrix} 1 & 0 \\ 0 & 1 \end{bmatrix} + \begin{bmatrix} 0 & 1 \\ -2 & -3 \end{bmatrix} t + \begin{bmatrix} 0 & 1 \\ -2 & -3 \end{bmatrix}^2 \frac{t^2}{2!} + \begin{bmatrix} 0 & 1 \\ -2 & -3 \end{bmatrix}^3 \frac{t^3}{3!} + \cdots$$

$$= \begin{bmatrix} 1 - t^2 + t^3 + \ldots & t - \frac{3}{2}t^2 - \frac{7}{6}t^3 + \ldots \\ -2t + 3t^2 - \frac{7}{3}t^3 + \ldots & 1 - 3t + \frac{7}{2}t^2 - \frac{5}{2}t^3 + \ldots \end{bmatrix}$$

Method 2: Diagonal Form

This method takes place entirely in the time domain and is based on the diagonalization of the matrix A. Indeed, if the eigenvalues of matrix A are distinct, then A can be transformed to a diagonal matrix Λ via the transformation matrix T as follows: $\Lambda = T^{-1}AT$. Matrix $\Phi(t)$, under the transformation T, becomes:

$$\Phi(t) = e^{At} = Te^{\Lambda t}T^{-1} .$$

Since:

$$e^{\Lambda t} = \begin{bmatrix} e^{\lambda_1 t} & & & 0 \\ & e^{\lambda_2 t} & & \\ & & \ddots & \\ 0 & & & e^{\lambda_n t} \end{bmatrix} , \quad \Lambda = \begin{bmatrix} \lambda_1 & & & 0 \\ & \lambda_2 & & \\ & & \ddots & \\ 0 & & & \lambda_n \end{bmatrix} ,$$

it follows that

$$\Phi(t) = e^{At} = T \begin{bmatrix} e^{\lambda_1 t} & & & 0 \\ & e^{\lambda_2 t} & & \\ & & \ddots & \\ 0 & & & e^{\lambda_n t} \end{bmatrix} T^{-1} .$$

Example 3.3. $A = \begin{bmatrix} 0 & 1 \\ -2 & -3 \end{bmatrix}$ is known. Now compute the matrix e^{At}.

Solution.

$$|\lambda I - A| = \begin{vmatrix} \lambda & -1 \\ 2 & \lambda + 3 \end{vmatrix} = \lambda^2 + 3\lambda + 2 = (\lambda + 1)(\lambda + 2) = 0,$$

so, $\lambda_1 = -1$, $\lambda_2 = -2$.

We can, therefore, have the corresponding transfer matrix according to equation (3.25).

$$T = \begin{bmatrix} 2 & 1 \\ -2 & -2 \end{bmatrix} \quad \text{and} \quad T^{-1} = \begin{bmatrix} 1 & \frac{1}{2} \\ -1 & -1 \end{bmatrix}.$$

Therefore,

$$e^{At} = \begin{bmatrix} 2 & 1 \\ -2 & -2 \end{bmatrix} \begin{bmatrix} e^{-t} & 0 \\ 0 & e^{-2t} \end{bmatrix} \begin{bmatrix} 1 & \frac{1}{2} \\ -1 & -1 \end{bmatrix}$$

$$= \begin{bmatrix} 2e^{-t} - e^{-2t} & e^{-t} - e^{-2t} \\ -2e^{-t} + 2e^{-2t} & -e^{-t} + 2e^{-2t} \end{bmatrix}.$$

If the matrix A has repeated eigenvalues, then A can be transformed to a Jordan matrix Λ via the transformation matrix T:

$$J = T^{-1}AT,$$

$$e^{At} = Te^{Jt}T^{-1}.$$

Example 3.4. Compute the matrix e^{At}, where $A = \begin{bmatrix} 0 & 1 & 0 \\ 0 & 0 & 1 \\ 2 & -5 & 4 \end{bmatrix}$.

Solution.

$$|\lambda I - A| = \begin{vmatrix} \lambda & -1 & 0 \\ 0 & \lambda & -1 \\ -2 & 5 & \lambda - 4 \end{vmatrix} = (\lambda - 1)^2 (\lambda - 2) = 0,$$

so,

$$\lambda_1 = \lambda_2 = 1, \quad \lambda_3 = 2.$$

According to (3.26),

$$J = \begin{bmatrix} 1 & 1 & 0 \\ 0 & 1 & 0 \\ 0 & 0 & 2 \end{bmatrix},$$

$$e^{Jt} = \begin{bmatrix} e^t & te^t & 0 \\ 0 & e^t & 0 \\ 0 & 0 & e^{2t} \end{bmatrix}.$$

Since

$$
T = \begin{bmatrix} 1 & -1 & 1 \\ 1 & 0 & 2 \\ 1 & 1 & 4 \end{bmatrix} ; \quad T^{-1} = \begin{bmatrix} -2 & 5 & -2 \\ -2 & 3 & -1 \\ 1 & -2 & 1 \end{bmatrix} .
$$

Then,

$$
\begin{aligned}
e^{At} &= \begin{bmatrix} 1 & -1 & 1 \\ 1 & 0 & 2 \\ 1 & 1 & 4 \end{bmatrix} \begin{bmatrix} e^t & te^t & 0 \\ 0 & e^t & 0 \\ 0 & 0 & e^{2t} \end{bmatrix} \begin{bmatrix} -2 & 5 & -2 \\ -2 & 3 & -1 \\ 1 & -2 & 1 \end{bmatrix} \\
&= \begin{bmatrix} e^t & te^t - e^t & e^{2t} \\ e^t & te^t & 2e^{2t} \\ e^t & te^t + e^t & 4e^{2t} \end{bmatrix} \begin{bmatrix} -2 & 5 & -2 \\ -2 & 3 & -1 \\ 1 & -2 & 1 \end{bmatrix} \\
&= \begin{bmatrix} -2te^t + e^{2t} & 3te^t + 2e^t - e^{2t} & -te^t - e^t + e^{2t} \\ 2(e^{2t} - te^t - e^t) & 3te^t + 5e^t - 4e^{2t} & -te^t - 2e^t + 2e^{2t} \\ -2te^t - 4e^t + 4e^{2t} & 3te^t + 8e^t - 8e^{2t} & -te^t - 3e^t + 4e^{2t} \end{bmatrix}
\end{aligned}
$$

Method 3: The Inverse Laplace Transformation

The inverse Laplace transformation method can be expressed as:

$$
e^{At} = \Phi(t) = \ell^{-1} \left\{ (sI - A)^{-1} \right\} . \tag{3.30}
$$

Proof. Consider the differential equation:

$$
\dot{x}(t) = Ax(t) ,
$$

with the initial state $x(0) = x_0$.

Apply the Laplace transform on both sides of the equation:

$$
sX(s) - x(0) = AX(s) .
$$

In other words:

$$
(sI - A)X(s) = x(0) = x_0 ,
$$

and, therefore:

$$
X(s) = (sI - A)^{-1} x_0 .
$$

Apply the inverse Laplace transform on both sides to get the solution of the differential equation:

$$
x(t) = \ell^{-1} \left\{ (sI - A)^{-1} \right\} x_0 .
$$

Comparing the above equation with equation (3.15) can result in:

$$
e^{At} = \Phi(t) = \ell^{-1} \left\{ (sI - A)^{-1} \right\} . \qquad \square
$$

Example 3.5. Compute the matrix e^{At}, where $A = \begin{bmatrix} 0 & 1 \\ -2 & -3 \end{bmatrix}$.

Solution.

$$sI - A = \begin{bmatrix} s & -1 \\ 2 & s+3 \end{bmatrix}$$

$$(sI - A)^{-1} = \frac{1}{|sI - A|} \text{adj}(sI - A) = \frac{1}{(s+1)(s+2)} \begin{bmatrix} s+3 & 1 \\ -2 & s \end{bmatrix}$$

$$= \begin{bmatrix} \frac{s+3}{(s+1)(s+2)} & \frac{1}{(s+1)(s+2)} \\ \frac{-2}{(s+1)(s+2)} & \frac{s}{(s+1)(s+2)} \end{bmatrix}$$

$$= \begin{bmatrix} \frac{2}{s+1} - \frac{1}{s+2} & \frac{1}{s+1} - \frac{1}{s+2} \\ \frac{-2}{s+1} + \frac{2}{s+2} & \frac{-1}{s+1} + \frac{2}{s+2} \end{bmatrix}$$

Therefore,

$$e^{At} = \ell^{-1}\left\{(sI - A)^{-1}\right\} = \begin{bmatrix} 2e^{-t} - e^{-2t} & e^{-t} - e^{-2t} \\ -2e^{-t} + 2e^{-2t} & -e^{-t} + 2e^{-2t} \end{bmatrix}.$$

Method 4: Cayley–Hamilton Theorem

(1) A square matrix A satisfies the characteristic equation of itself according to the Cayley–Hamilton theorem:

$$f(A) = A^n + a_{n-1}A^{n-1} + \cdots + a_1 A + a_0 I = 0,$$

thus,

$$A^n = -a_{n-1}A^{n-1} - a_{n-2}A^{n-2} - \cdots - a_1 A - a_0 I,$$

which is the linear combination of $A^{n-1}, A^{n-2}, \ldots, A, I$.

In the same way,

$$A^{n+1} = A \cdot A^n = -a_{n-1}A^n - (a_{n-2}A^{n-1} + a_{n-3}A^{n-2} + \cdots + a_1 A^2 + a_0 A)$$

$$= -a_{n-1}\left(-a_{n-1}A^{n-1} - a_{n-2}A^{n-2} - \cdots - a_1 A - a_0 I\right)$$

$$- (a_{n-2}A^{n-1} + a_{n-3}A^{n-2} + \cdots + a_1 A^2 + a_0 A)$$

$$= \left(a_{n-1}^2 - a_{n-2}\right)A^{n-1} + (a_{n-1}a_{n-2} - a_{n-3})A^{n-2} + \cdots$$

$$+ (a_{n-1}a_1 - a_0)A + a_{n-1}a_0 I.$$

That is to say, $A^n, A^{n+1} \ldots$ can all be expressed with $A^{n-1}, A^{n-2}, \ldots, A, I$.

(2) In the definition equation (3.29) of e^{At}, we can eliminate the terms of A with power equal to and above n, applying the method in (1). In other words:

$$e^{At} = I + At + \frac{1}{2!}A^2 t^2 + \cdots + \frac{1}{(n-1)!}A^{n-1}t^{n-1} + \frac{1}{n!}A^n t^n + \frac{1}{(n+1)!}A^{n+1}t^{n+1} + \cdots$$

$$= a_{n-1}(t)A^{n-1} + a_{n-2}(t)A^{n-2} + \cdots + a_1(t)A + a_0(t)I. \tag{3.31}$$

Example 3.6. Compute $a_i(t)$ in the expression of e^{At}, where $A = \begin{bmatrix} 0 & 1 \\ -2 & -3 \end{bmatrix}$.

Solution. The characteristic equation of A:

$$|\lambda I - A| = \begin{vmatrix} \lambda & -1 \\ 2 & \lambda + 3 \end{vmatrix} = \lambda^2 + 3\lambda + 2 = 0 .$$

According to the Cayley–Hamilton theorem,

$$A^2 + 3A + 2I = 0 .$$

Thus:

$$A^2 = -3A - 2I .$$

While

$$A^3 = A \cdot A^2 = A(-3A - 2I) = -3A^2 - 2A$$
$$= -3(-3A - 2I) - 2A = 7A - 6I ,$$

$$A^4 = A \cdot A^2 = 7A^2 + 6A$$
$$= 7(-3A - 2I) + 6A = -15A - 14I .$$

$$\cdots$$

Substituting the equations above into the following equation, we can eliminate the terms of A with power equal to and above two:

$$e^{At} = I + At + \frac{1}{2!}A^2 t^2 + \frac{1}{3!}A^3 t^3 + \frac{1}{4!}A^4 t^4 + \cdots$$
$$= \left(t - \frac{3}{2!}t^2 + \frac{7}{3!}t^3 - \frac{15}{4!}t^4 + \cdots \right) A + \left(1 - t^2 + t^3 - \frac{14}{4!}t^4 + \cdots \right) I$$
$$= a_1(t)A + a_0(t)I .$$

Therefore,

$$a_1(t) = t - \frac{3}{2!}t^2 + \frac{7}{3!}t^3 - \frac{15}{4!}t^4 + \cdots$$
$$a_0(t) = 1 - t^2 + t^3 - \frac{14}{4!}t^4 + \cdots .$$

(3) When the eigenvalues of A are all distinct, we have:

$$\begin{bmatrix} a_0(t) \\ a_1(t) \\ \vdots \\ a_{n-1}(t) \end{bmatrix} = \begin{bmatrix} 1 & \lambda_1 & \lambda_1^2 & \cdots & \lambda_1^{n-1} \\ 1 & \lambda_2 & \lambda_2^2 & \cdots & \lambda_2^{n-1} \\ \vdots & & & & \\ 1 & \lambda_n & \lambda_n^2 & \cdots & \lambda_n^{n-1} \end{bmatrix}^{-1} \begin{bmatrix} e^{\lambda_1 t} \\ e^{\lambda_2 t} \\ \vdots \\ e^{\lambda_n t} \end{bmatrix} . \tag{3.32}$$

Proof. Matrix A satisfies the characteristic equation of itself and, therefore, eigenvalues λ and A are exchangeable. Thus, λ satisfies equation (3.31):

$$\left.\begin{array}{c} a_0(t) + a_1(t)\lambda_1 + \cdots + a_{n-1}(t)\lambda_1^{n-1} = e^{\lambda_1 t} \\ a_0(t) + a_1(t)\lambda_2 + \cdots + a_{n-1}(t)\lambda_2^{n-1} = e^{\lambda_2 t} \\ \vdots \\ a_0(t) + a_1(t)\lambda_n + \cdots + a_{n-1}(t)\lambda_n^{n-1} = e^{\lambda_n t} \end{array}\right\}$$

Solve the above equation for $[a_0(t) \quad a_1(t) \quad \cdots \quad a_{n-1}(t)]^T$, and we can obtain (3.32). $\square$

When eigenvalues of A are all λ_1, we have:

$$\begin{bmatrix} a_0(t) \\ a_1(t) \\ \vdots \\ a_{n-2}(t) \\ a_{n-1}(t) \end{bmatrix} = \begin{bmatrix} 0 & 0 & 0 & \cdots & 0 & 1 \\ 0 & 0 & 0 & \cdots & 1 & (n-1)\lambda_1 \\ \vdots & \vdots & \vdots & & & \vdots \\ 0 & 0 & 1 & \cdots & & \frac{(n-1)(n-2)}{2!}\lambda_1^{n-3} \\ 0 & 1 & 2\lambda & \cdots & (n-1)\lambda_1^{n-2} & (n-1)\lambda_1^{n-2} \\ 1 & \lambda_1 & \lambda_1^2 & \cdots & \lambda_1^{n-1} & \lambda_1^{n-1} \end{bmatrix}^{-1} \begin{bmatrix} \frac{1}{(n-1)!}t^{n-1}e^{\lambda_1 t} \\ \frac{1}{(n-2)!}t^{n-2}e^{\lambda_1 t} \\ \vdots \\ \frac{1}{2!}t^2 e^{\lambda_1 t} \\ t e^{\lambda_1 t} \\ e^{\lambda_1 t} \end{bmatrix}.$$

$$(3.33)$$

Proof.

$$a_0(t) + a_1(t)\lambda_1 + a_2(t)\lambda_1^2 \cdots + a_{n-1}(t)\lambda_1^{n-1} = e^{\lambda_1 t}.$$

Differentiating both sides of the equation, we have:

$$a_1(t) + 2a_2(t)\lambda_1 \cdots + (n-1)a_{n-1}(t)\lambda_1^{n-2} = t e^{\lambda_1 t}.$$

Differentiating both sides of the equation again, we obtain:

$$2a_2(t) + 6a_3(t) + \cdots + (n-1)(n-2)a_{n-1}(t)\lambda_1^{n-3} = t^2 e^{\lambda_1 t}.$$

Repeat the former step and, finally, we have:

$$(n-1)!a_{n-1}(t) = t^{n-1}e^{\lambda_1 t}.$$

The equation (3.33) can be reached from the above n equations, by solving $a_i(t)$. $\square$

Example 3.7. Compute e^{At}, where $A = \begin{bmatrix} 0 & 1 \\ -2 & -3 \end{bmatrix}$.

Solution. We know the eigenvalues of the matrix A is $\lambda_1 = -1$, $\lambda_2 = -2$. According to equation (3.33), we have:

$$\begin{bmatrix} a_0 \\ a_1 \end{bmatrix} = \begin{bmatrix} 1 & \lambda_1 \\ 1 & \lambda_2 \end{bmatrix}^{-1} \begin{bmatrix} e^{\lambda_1 t} \\ e^{\lambda_2 t} \end{bmatrix} = \begin{bmatrix} 1 & -1 \\ 1 & -2 \end{bmatrix}^{-1} \begin{bmatrix} e^{-t} \\ e^{-2t} \end{bmatrix}$$

$$= \begin{bmatrix} 2 & -1 \\ 1 & -1 \end{bmatrix} \begin{bmatrix} e^{-t} \\ e^{-2t} \end{bmatrix} = \begin{bmatrix} 2e^{-t} - e^{-2t} \\ e^{-t} - e^{-2t} \end{bmatrix}.$$

Therefore,

$$e^{At} = a_0(t)I + a_1(t)A$$

$$= (2e^{-t} - e^{-2t})\begin{bmatrix} 1 & 0 \\ 0 & 1 \end{bmatrix} + (e^{-t} - e^{-2t})\begin{bmatrix} 0 & 1 \\ -2 & -3 \end{bmatrix}$$

$$= \begin{bmatrix} 2e^{-t} - e^{-2t} & e^{-t} - e^{-2t} \\ -2e^{-t} + 2e^{-2t} & -e^{-t} + 2e^{-2t} \end{bmatrix}.$$

Example 3.8. Compute the matrix e^{At}, where $A = \begin{bmatrix} 0 & 1 & 0 \\ 0 & 0 & 1 \\ 2 & -5 & 4 \end{bmatrix}$.

Solution. The eigenvalues of the matrix A are $\lambda_1 = \lambda_2 = 1, \lambda_3 = 2$. The part of $\lambda_1 = \lambda_2 = 1$ can be calculated based on equation (3.33), while the part of $\lambda_3 = 2$ can be calculated based on equation (3.32):

$$\begin{bmatrix} a_0 \\ a_1 \\ a_2 \end{bmatrix} = \begin{bmatrix} 0 & 1 & 2\lambda_1 \\ 1 & \lambda_1 & \lambda_1^2 \\ 1 & \lambda_3 & \lambda_3^2 \end{bmatrix}^{-1} \begin{bmatrix} te^{\lambda_1 t} \\ e^{\lambda_1 t} \\ e^{\lambda_3 t} \end{bmatrix}$$

$$= \begin{bmatrix} 0 & 1 & 2 \\ 1 & 1 & 1 \\ 1 & 2 & 4 \end{bmatrix}^{-1} \begin{bmatrix} te^t \\ e^t \\ e^{2t} \end{bmatrix} = \begin{bmatrix} -2 & 0 & 1 \\ 3 & 2 & -2 \\ -1 & -1 & 1 \end{bmatrix} \begin{bmatrix} te^t \\ e^t \\ e^{2t} \end{bmatrix}.$$

Therefore,

$$e^{At} = (-2te^t + e^{2t})\begin{bmatrix} 1 & 0 & 0 \\ 0 & 1 & 0 \\ 0 & 0 & 1 \end{bmatrix} + (3te^t + 2e^t - 2e^{2t})\begin{bmatrix} 0 & 1 & 0 \\ 0 & 0 & 1 \\ 2 & -5 & 4 \end{bmatrix}$$

$$+ (-te^t - e^t + e^{2t})\begin{bmatrix} 0 & 0 & 1 \\ 2 & -5 & 4 \\ 8 & -18 & 11 \end{bmatrix}$$

$$= \begin{bmatrix} -2te^t + e^{2t} & 3te^t + 2e^t - 2e^{2t} & -te^t - e^t + e^{2t} \\ 2(e^{2t} - te^t - e^t) & 3te^t + 5e^t - 4e^{2t} & -te^t - 2e^t + 2e^{2t} \\ -2te^t - 4e^t + 4e^{2t} & 3te^t + 8e^t - 8e^{2t} & -te^t - 3e^t + 4e^{2t} \end{bmatrix}.$$

Example 3.9. Consider the state equation:

$$\dot{x} = \begin{bmatrix} -2 & 0 & 0 \\ 1 & 0 & 1 \\ 0 & -2 & -2 \end{bmatrix} x + \begin{bmatrix} 1 \\ 0 \\ 1 \end{bmatrix} u$$

$$y = \begin{bmatrix} 1 & -1 & 0 \end{bmatrix} x.$$

Suppose the input is a step function of various magnitudes. First we use MATLAB to find its unit step response. We type:

```
a=[-2 0 0;1 0 1;0 -2 -2];
b=[1;0;1];
c=[0 -2 -2];
d=0;
[y,x,t]=step(a,b,c,d);
plot(t,y,t,x)
```

The system response is shown in Figure 3.2.

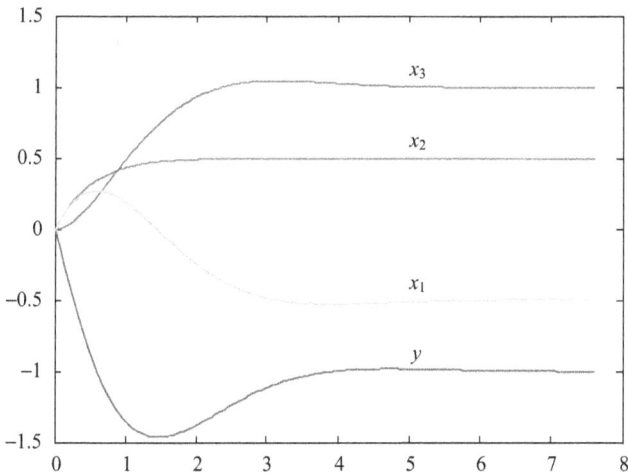

Fig. 3.2: State variables of system step response.

3.4 Discretization

Consider the continuous time state space equation:

$$\dot{x}(t) = Ax(t) + Bu(t) . \tag{3.34}$$

If this continuous time equation is to be computed on a digital computer, it must be discretized, because:

$$\dot{x} = \lim_{T \to 0} \frac{x(t + T) - x(t)}{T} .$$

When the sample period T is quite small, about $1/10$ of the minimum time constant of the system, (3.34) can be approximated as:

$$x[(k + 1)T] = (TA + 1)x(kT) + TBu(kT) . \tag{3.35}$$

Proof. According to the definition of derivative, we have:

$$\dot{x}(t_0) = \lim_{\Delta t \to 0} \frac{x(t_0 + \Delta t) - x(t_0)}{\Delta t} . \tag{3.36}$$

If we compute $x(t)$ and $y(t)$ from $t_0 = kT$ to $t = (k+1)T$ for $k = 0, 1, \ldots$, then (3.36) becomes:

$$\dot{x}(kT) = \lim_{T \to 0} \frac{x[(k+1)T] - x(kT)}{T} \approx \frac{x[(k+1)T] - x(kT)}{T} . \tag{3.37}$$

Substituting (3.37) into (3.34) yields:

$$\frac{x[(k+1)T] - x(kT)}{T} = Ax(kT) + Bu(kT) ,$$

which can be rearranged as (3.35), thus providing proof. □

This is a discrete time state space equation and can easily be computed on a digital computer. This discretization is the easiest to carry out but yields the least accurate results. The following is an alternative discretization method.

If an input $u(t)$ is generated by a digital computer followed by a digital to analog converter, then $u(t)$ will be piecewise constant. This situation often arises in computer control of control systems. Let

$$u(t) = u(kT) = \text{constant} \quad \text{for} \quad kT \le t < (k+1)T$$

for $k = 0, 1, 2, \ldots$. This input changes value only at discrete time instants. For this input, the solution of the state equation in (3.34) still equals (3.7). Computing (3.7) at $t = kT$ and $t = (k+1)T$ yields:

$$x[(k+1)T] = e^{AT}x(kT) + \int_{kT}^{(k+1)T} e^{A[(k+1)T-\tau]}Bd\tau u(kT) . \tag{3.38}$$

Let $t = (k+1)T - \tau$, then $d\tau = -dt$. So the lower integral $\tau = kT$ becomes $t = T$; and the higher integral $\tau = (k+)T$ becomes $t = 0$. Thus, equation (3.38) can be simplified as:

$$x[(k+1)T] = e^{AT}x(kT) + \int_0^T e^{At}dtBu(kT) ,$$

which equals

$$x(k+1) = e^{AT}x(k) + \int_0^T e^{At}dtBu(k) . \tag{3.39}$$

This is a discrete time state space equation. Note that there is no approximation involved in this derivation. It is the exact solution of (3.34) at $t = kT$, if the input is piecewise constant.

Rewrite (3.39) as:

$$x(k + 1) = G(T)x(k) + H(T)u(k) ,\tag{3.40}$$

where

$$G(T) = e^{AT} , \quad H(T) = \int_0^T e^{At}dt \cdot B .\tag{3.41}$$

The MATLAB function [ad,bd]=c2d(a,b,T) transforms the continuous state equation in (3.34) into the discrete time state equation in (3.41).

Example 3.10. Try to discretize the following state equation:

$$\dot{x} = \begin{bmatrix} 0 & 1 \\ 0 & -2 \end{bmatrix} x(t) + \begin{bmatrix} 0 \\ 1 \end{bmatrix} u(t) .$$

Solution. Step 1. Equation (3.41) yields the exact result:

$$e^{At} = \ell^{-1}\left[(sI - A)^{-1}\right] = \ell^{-1}\left\{\begin{bmatrix} s & -1 \\ 0 & s+2 \end{bmatrix}^{-1}\right\} = \begin{bmatrix} 1 & \frac{1}{2}(1 - e^{-2T}) \\ 0 & e^{-2T} \end{bmatrix} .$$

Thus, we have:

$$G(T) = \begin{bmatrix} 1 & \frac{1}{2}(1 - e^{-2T}) \\ 0 & e^{-2T} \end{bmatrix}$$

$$H = \int_0^T e^{At}dt \cdot B = \int_0^T \begin{bmatrix} 1 & \frac{1}{2}(1 - e^{-2t}) \\ 0 & e^{-2t} \end{bmatrix} dt \cdot \begin{bmatrix} 0 \\ 1 \end{bmatrix}$$

$$= \begin{bmatrix} T & \frac{1}{2}\left(T + \frac{1}{2}e^{-2T} - \frac{1}{2}\right) \\ 0 & -\frac{1}{2}e^{-2T} + \frac{1}{2} \end{bmatrix} \begin{bmatrix} 0 \\ 1 \end{bmatrix} = \begin{bmatrix} \frac{1}{2}\left(T + \frac{e^{-2T}-1}{2}\right) \\ \frac{1}{2}(1 - e^{-2T}) \end{bmatrix} .$$

According to equation (3.40), the discretized state equation is:

$$x(k + 1) = \begin{bmatrix} 1 & \frac{1}{2}(1 - e^{-2T}) \\ 0 & e^{-2T} \end{bmatrix} x(k) + \begin{bmatrix} \frac{1}{2}\left(T + \frac{e^{-2T}-1}{2}\right) \\ \frac{1}{2}(1 - e^{-2T}) \end{bmatrix} u(k) .$$

Step 2. Equation (3.35) yields the approximate result:

$$TA + I = \begin{bmatrix} 0 & T \\ 0 & -2T \end{bmatrix} + \begin{bmatrix} 1 & 0 \\ 0 & 1 \end{bmatrix} = \begin{bmatrix} 1 & T \\ 0 & 1 - 2T \end{bmatrix}$$

$$H = TB = \begin{bmatrix} 0 \\ T \end{bmatrix} .$$

According to equation (3.45), the discretized state equation is:

$$x[(k + 1)T] = \begin{bmatrix} 1 & T \\ 0 & 1 - 2T \end{bmatrix} x(kT) + \begin{bmatrix} 0 \\ T \end{bmatrix} u(kT) ,$$

which can be rewritten as:

$$x(k) = \begin{bmatrix} 1 & T \\ 0 & 1-2T \end{bmatrix} x(k) + \begin{bmatrix} 0 \\ T \end{bmatrix} u(k) .$$

Then we use MATLAB to finish this task. We type:

```
a=[0 1;0 -2];
b=[0;1];
T=0.1;
[ad,bd]=c2d(a,b,T)
```

Yield

```
ad =
    1.0000    0.0906
    0         0.8187
bd =
    0.0047
    0.0906
```

We can get:

$$x(k) = \begin{bmatrix} 1 & 0.09 \\ 0 & 0.8 \end{bmatrix} x(k) + \begin{bmatrix} 0.0047 \\ 0.09 \end{bmatrix} u(k) .$$

3.5 Solution of Discrete Time Equation

Method 1: Recursive Method

Consider the discrete time state space equation:

$$x(k+1) = G(T)x(k) + H(T)u(k)$$
$$x(k)|_{k=0} = x(0) . \tag{3.42}$$

The solution of the first order matrix differential equation is:

$$x(k) = G^k x(0) + \sum_{j=0}^{k-1} G^{k-j-1} Hu(j) . \tag{3.43a}$$

Or:

$$x(k) = G^k x(0) + \sum_{j=0}^{k-1} G^j Hu(k-j-1) , \tag{3.43b}$$

which equals:

$$x(k) = G^k x(0) + G^{k-1} Hu(0) + G^{k-2} Hu(1) + \cdots + GHu(k-2) + Hu(k-1) . \tag{3.43c}$$

Proof. Solve the matrix differential equation (3.42) by iterative method:

For $i = 0$, $x(1) = Gx(0) + Hu(0)$.

For $i = 1$, $x(2) = Gx(1) + Hu(1) = G^2x(0) + GHu(1) + Hu(1)$.

For $i = 2$, $x(3) = Gx(2) + Hu(2) = G^3x(0) + G^2Hu(0) + GHu(1) + Hu(2)$.

$$\vdots$$

For $i = k - 1$, $x(k) = Gx(k - 1) + Hu(k - 1) = G^kx(0) + G^{k-1}Hu(0) + \cdots$
$$+ GHu(k - 2) + Hu(k - 1).$$

□

The last general formula is just equation (3.43c).

Equation (3.43c) can be expressed in matrix form as:

$$
\begin{bmatrix} x(1) \\ x(2) \\ x(3) \\ \vdots \\ x(k) \end{bmatrix} = \begin{bmatrix} G \\ G^2 \\ G^3 \\ \vdots \\ G^k \end{bmatrix} x(0) + \begin{bmatrix} H & 0 & 0 & \cdots & 0 \\ GH & H & 0 & \cdots & 0 \\ G^2H & GH & H & \cdots & 0 \\ \vdots & \vdots & \vdots & & \vdots \\ G^{k-1}H & G^{k-2}H & G^{k-3}H & \cdots & H \end{bmatrix}.
\tag{3.43d}
$$

Solution (3.43) is derived from the initial time instant $k = 0$. If we start from the time $k = h$, the corresponding initial state is $x(h)$. Then the solution becomes:

$$x(k) = G^{k-h}x(0) + \sum_{j=h}^{k-1} G^{k-j-1}Hu(j), \tag{3.44a}$$

or:

$$x(k) = G^{k-h}x(0) + \sum_{j=h}^{k-1} G^jHu(k - j - 1). \tag{3.44b}$$

Obviously, the solution of the discrete time state space equation is similar to that of the continuous time state space equation. It consists of two parts of responses; the response excited by the initial state and the response excited by the input signal. Furthermore, the solution of the discrete time state space equation is a discrete track in state space. Besides, in the response excited by the input, the state $x(k)$ is only related with the sample values of input before time instant k.

Similarly, we define

$$\Phi(k) = G^k \quad \text{or} \quad \Phi(k - h) = G^{k-h} \tag{3.45}$$

as the state transition matrix of the discrete time system. Obviously,

$$\Phi(k + 1) = G\Phi(k); \quad \Phi(0) = I \tag{3.46}$$

and the following properties hold:

$$\Phi(k - h) = \Phi(k - h_1)\Phi(h_1 - h) \quad \text{for} \quad k > h_1 \geq h \tag{3.47}$$

$$\Phi^{-1}(k) = \Phi(-k) . \tag{3.48}$$

Using the state transition matrix $\Phi(k)$, the solution (3.43) can be expressed as:

$$x(k) = \Phi(k)x(0) + \sum_{j=0}^{k-1} \Phi(k - j - 1)Hu(j) , \tag{3.49a}$$

or as:

$$x(k) = \Phi(k)x(0) + \sum_{j=0}^{k-1} \Phi(k - j - 1)Hu(j). \tag{3.49b}$$

Thus, equation (3.44) can be written as:

$$x(k) = \Phi(k - h)x(h) + \sum_{j=h}^{k-1} \Phi(k - j - 1)Hu(j) \tag{3.50a}$$

$$x(k) = \Phi(k - h)x(0) + \sum_{j=h}^{k-1} \Phi(j)Hu(k - j - 1) . \tag{3.50b}$$

Example 3.11. The state equation of a discrete time system is:

$$x(k + 1) = Gx(k) + Hu(k)$$

$$G = \begin{bmatrix} 0 & 1 \\ -0.16 & -1 \end{bmatrix} , \quad H = \begin{bmatrix} 1 \\ 1 \end{bmatrix} ,$$

with initial state $x(0) = [1 \quad -1]^T$ and control action $u(k) = 1$. Try to solve $\Phi(k)$, $x(k)$.

Solution. As defined.

$$\Phi(k) = G^k = \begin{bmatrix} 0 & 1 \\ -0.16 & -1 \end{bmatrix}^k .$$

For simplicity, we transform the original equation into Jordan canonical form, i.e., transform G into diagonal form.

Let $x(k) = T\tilde{x}(k)$. Thus, the original equation becomes:

$$\tilde{x}(k + 1) = T^{-1}GT\tilde{x}(k) + T^{-1}Hu(k) .$$

Again, let

$$T^{-1}GT = \Lambda ; \quad \tilde{\Phi}(k) = (T^{-1}GT)^k = \Lambda^k ,$$

so

$$\tilde{x}(k) = \tilde{\Phi}(k)\tilde{x}(0) + \sum_{j=0}^{k-1} \tilde{\Phi}(j)T^{-1}Hu(k - j - 1) \tag{3.51}$$

$$|\lambda I - G| = \begin{vmatrix} \lambda & -1 \\ 0.16 & \lambda + 1 \end{vmatrix} = (\lambda + 0.2)(\lambda + 0.8) = 0$$

$$\lambda_1 = -0.2 ; \quad \lambda_2 = -0.8 .$$

Therefore,

$$\Lambda = \begin{bmatrix} -0.2 & 0 \\ 0 & -0.8 \end{bmatrix} ; \quad \widetilde{\Phi}(k) = \begin{bmatrix} -0.2 & 0 \\ 0 & -0.8 \end{bmatrix}^k = \begin{bmatrix} (-0.2)^k & 0 \\ 0 & (-0.8)^k \end{bmatrix},$$

thus

$$T = \begin{bmatrix} 1 & 1 \\ -0.2 & -0.8 \end{bmatrix}, \quad T^{-1} = \begin{bmatrix} \frac{4}{3} & \frac{5}{3} \\ -\frac{1}{3} & -\frac{5}{3} \end{bmatrix}.$$

Now, it is easy to derive

$$\Phi(k) = T\widetilde{\Phi}(k)T^{-1} = \begin{bmatrix} 1 & 1 \\ -0.2 & -0.8 \end{bmatrix} \begin{bmatrix} (-0.2)^k & 0 \\ 0 & (-0.8)^k \end{bmatrix} \begin{bmatrix} \frac{4}{3} & \frac{5}{3} \\ -\frac{1}{3} & -\frac{5}{3} \end{bmatrix}$$

$$= \frac{1}{3} \begin{bmatrix} 4(-0.2)^k - (-0.8)^k & 5\left[(-0.2)^k - (-0.8)^k\right] \\ -0.8\left[(-0.2)^k - (-0.8)^k\right] & -(-0.2)^k + 4(-0.8)^k \end{bmatrix}.$$

Now compute $\tilde{x}(k)$ according to equation (3.51). The first term on the right side is:

$$\Phi(k)\tilde{x}(0) = \Phi(k)T^{-1}x(0) = \begin{bmatrix} (-0.2)^k & 0 \\ 0 & (-0.8)^k \end{bmatrix} \begin{bmatrix} \frac{4}{3} & \frac{5}{3} \\ -\frac{1}{3} & -\frac{5}{3} \end{bmatrix} \begin{bmatrix} 1 \\ -1 \end{bmatrix} = \frac{1}{3} \begin{bmatrix} -(-0.2)^k \\ 4(-0.8)^k \end{bmatrix}.$$

The second term on the right side is:

$$\sum_{j=0}^{k-1} \Phi(j)T^{-1}Hu(k-j-1) = \sum_{j=0}^{k-1} \Phi(j) \begin{bmatrix} \frac{4}{3} & \frac{5}{3} \\ -\frac{1}{3} & -\frac{5}{3} \end{bmatrix} \begin{bmatrix} 1 \\ 1 \end{bmatrix} [1]$$

$$= \sum_{j=0}^{k-1} \begin{bmatrix} (-0.2)^j & 0 \\ 0 & (-0.8)^j \end{bmatrix} \begin{bmatrix} 3 \\ -2 \end{bmatrix} = \sum_{j=0}^{k-1} \begin{bmatrix} 3(-0.2)^j \\ -2(-0.8)^j \end{bmatrix}$$

$$= \begin{bmatrix} 3\left[1 + (-0.2) + (-0.2)^2 + \cdots + (-0.2)^{k-1}\right] \\ -2\left[1 + (-0.8) + (-0.8)^2 + \cdots + (-0.8)^{k-1}\right] \end{bmatrix}$$

$$= \begin{bmatrix} \frac{3[1-(-0.2)^k]}{1.2} \\ \frac{-2[1-(-0.8)^k]}{1.8} \end{bmatrix}.$$

Thus:

$$\tilde{x}(k) = \frac{1}{3} \begin{bmatrix} -(-0.2)^k \\ 4(-0.8)^k \end{bmatrix} + \begin{bmatrix} \frac{1}{0.4}\left[1 - (-0.2)^k\right] \\ -\frac{1}{0.9}\left[1 - (-0.8)^k\right] \end{bmatrix}$$

$$= \begin{bmatrix} -\frac{17}{6}(-0.2)^k + \frac{5}{2} \\ \frac{22}{9}(-0.8)^k - \frac{10}{9} \end{bmatrix}.$$

Therefore:

$$x(k) = T\tilde{x}(k) = \begin{bmatrix} 1 & 1 \\ -0.2 & -0.8 \end{bmatrix} \begin{bmatrix} -\frac{17}{6}(-0.2)^k + \frac{5}{2} \\ \frac{22}{9}(-0.8)^k - \frac{10}{9} \end{bmatrix}$$

$$= \begin{bmatrix} -\frac{17}{6}(-0.2)^k + \frac{22}{9}(-0.8)^k + \frac{25}{18} \\ \frac{3.4}{6}(-0.2)^k - \frac{17.6}{9}(-0.8)^k + \frac{7}{18} \end{bmatrix}.$$

Method 2: z Transform Method

For the LTI discrete system state equation, we can find its solution by using the z transform method.

Consider the discrete time state space equation:

$$x(k + 1) = G(T)x(k) + H(T)u(k) .$$

Applying z transform to the above equation yields:

$$zx(z) - zx(0) = Gx(z) + Hu(z) ,$$

or:

$$(zI - G)x(z) = zx(0) + Hu(z) .$$

Thus,

$$x(z) = (zI - G)^{-1}zx(0) + (zI - G)^{-1}Hu(z) .$$

Take inverse z transform:

$$x(k) = \ell^{-1}\left[(zI - G)^{-1}zx(0)\right] + \ell^{-1}\left[(zI - G)^{-1}Hu(z)\right] . \tag{3.52}$$

Comparing (3.43) with (3.52) yields:

$$G^k x(0) = \ell^{-1}\left[(zI - G)^{-1}zx(0)\right] \tag{3.53}$$

$$\sum_{j=0}^{k-1} G^{k-j-1}Hu(j) = \ell^{-1}\left[(zI - G)^{-1}Hu(z)\right] . \tag{3.54}$$

Using the solution of the continuous state equation, we get:

$$x(t) = \Phi(t - kT)x(kT) + \int_{kT}^{t} \Phi(t - \tau)Bu(kT)d\tau .$$

Suppose that $t = (k + \Delta)T$ ($0 \le \Delta \le 1$) – the above equation becomes:

$$x[(k + \Delta)T] = \Phi(\Delta T)x(kT) + \int_{0}^{\Delta T} \Phi(\Delta T - \tau)d\tau Bu(kT) . \tag{3.55}$$

Comparing (3.43) with (3.52) yields:

$$G^k = \Phi(k) = \ell^{-1}\left[(zI - G)^{-1}z\right] \tag{3.56}$$

$$\sum_{j=0}^{k-1} G^{k-j-1}Hu(j) = \ell^{-1}\left[(zI - G)^{-1}Hu(z)\right] . \tag{3.57}$$

Proof. First we compute the z transform of G^k:

$$\ell[G^k] = \sum_{k=0}^{\infty} G^k z^{-k} = I + Gz^{-1} + G^2 z^{-2} + \dots . \tag{3.58}$$

Then we premultiply Gz^{-1} to both sides of (3.58):

$$Gz^{-1}\ell[G^k] = Gz^{-1} + G^2 z^{-12} + G^3 z^{-3} + \dots . \tag{3.59}$$

Subtract (3.58) from (3.59):

$$(I - Gz^{-1})\ell[G^k] = I ,$$

because

$$\ell[G^k] = \left(I - Gz^{-1}\right)^{-1} = (zI - G)^{-1}z . \tag{3.60}$$

Take z inverse transform of equation (3.60), and we can get equation (3.56).

Next we use convolution formula to prove equation (3.57):

$$\ell\left[\sum_{j=0}^{k-1} G^{k-j-1}Hu(j)\right] = \ell[G^{k-1}]H\ell[u(k)]$$

$$= \ell[G^k]z^{-1}H\ell[u(k)] = (zI - G)^{-1}Hu(z) .$$

Take z inverse transform of the above equation, then equation (3.57) is derived.

$$\sum_{j=0}^{k-1} G^{k-j-1}Hu(j) = \ell^{-1}\left[(zI - G)^{-1}Hu(z)\right] . \qquad \Box$$

Example 3.12. Consider the state equation in Example 3.11; try to find $\Phi(k)$ and $x(k)$ using the z transform method.

Solution. As $u(k) = 1$, we have:

$$u(z) = \frac{z}{z - 1} .$$

According to equation (3.56),

$$\Phi(k) = \ell^{-1}\left[(zI - G)^{-1}z\right]$$

$$= \ell^{-1}\left\{\begin{bmatrix} z & -1 \\ 0.16 & z+1 \end{bmatrix}^{-1} z\right\} = \ell^{-1}\left\{\frac{z}{(z+0.2)(z+0.8)}\begin{bmatrix} z+1 & 1 \\ -0.16 & z \end{bmatrix}\right\}$$

$$= \ell^{-1}\left\{\frac{z}{3}\begin{bmatrix} \frac{4}{z+0.2} + \frac{-1}{z+0.8} & \frac{5}{z+0.2} + \frac{-5}{z+0.8} \\ \frac{-0.8}{z+0.2} + \frac{0.8}{z+0.8} & \frac{-1}{z+0.2} + \frac{4}{z+0.8} \end{bmatrix}\right\}$$

$$= \frac{1}{3}\begin{bmatrix} 4(-0.2)^k - (-0.8)^k & 5(-0.2)^k - 5(-0.8)^k \\ -0.8(-0.2)^k + (-0.8)^k & -(-0.2)^k + 4(-0.8)^k \end{bmatrix} .$$

Now compute:

$$zx(0) + Hu(z) = \begin{bmatrix} z \\ -z \end{bmatrix} + \begin{bmatrix} \frac{z}{z-1} \\ \frac{z}{z-1} \end{bmatrix} = \begin{bmatrix} \frac{z^2}{z-1} \\ \frac{-z^2+2z}{z-1} \end{bmatrix}.$$

Thus,

$$x(z) = (zI - G)^{-1} [zx(0) + Hu(z)]$$

$$= \begin{bmatrix} \frac{(z^2+2)z}{(z+2)(z+0.8)(z-1)} \\ \frac{(-z^2+1.84z)z}{(z+2)(z+0.8)(z-1)} \end{bmatrix} = \begin{bmatrix} \frac{-(17/6)z}{z+2} + \frac{(22/9)z}{z+0.8} + \frac{(25/18)z}{z-1} \\ \frac{(3.4/6)z}{z+2} + \frac{(-17.6/9)z}{z+0.8} + \frac{(7/18)z}{z-1} \end{bmatrix}.$$

hence,

$$x(k) = \ell^{-1}[x(z)] = \begin{bmatrix} -\frac{17}{6}(-0.2)^k + \frac{22}{9}(-0.8)^k + \frac{25}{18} \\ \frac{3.4}{6}(-0.2)^k - \frac{17.6}{9}(-0.8)^k + \frac{7}{18} \end{bmatrix}.$$

3.6 Summary

The solution to both the continuous time and discrete time state space equation has been studied in this chapter. The state transfer matrix is a very important parameter matrix, which plays a big role in the solution of a state space equation. Different algorithms are discussed to obtain the state transfer matrix.

⚠ Exercise

3.1. Consider the matrix A:

$$A = \begin{pmatrix} 0 & 1 & 0 \\ 0 & 0 & 1 \\ 2 & -5 & 4 \end{pmatrix}.$$

Use the Laplace transform to find e^{-At}.

3.2. Use three different methods to find e^{-At}.

$$(1) \quad A = \begin{pmatrix} 0 & -1 \\ 4 & 0 \end{pmatrix}. \qquad (2) \quad A = \begin{pmatrix} 1 & 1 \\ 4 & 1 \end{pmatrix}.$$

3.3. Examine the following matrix to see whether they meet the conditions of state transition matrix. If they do, try to find out the corresponding matrix A.

$$(1) \qquad \Phi(t) = \begin{pmatrix} 1 & 0 & 0 \\ 0 & \sin t & \cos t \\ 0 & -\cos t & \sin t \end{pmatrix}.$$

$$(2) \qquad \Phi(t) = \begin{pmatrix} 1 & \frac{1}{2}(1 - e^{-2t}) \\ 0 & e^{-2t} \end{pmatrix}.$$

$$(3) \qquad \Phi(t) = \begin{pmatrix} 2e^{-t} - e^{-2t} & 2e^{-t} - 2e^{-2t} \\ e^{-t} - e^{-2t} & 2e^{-t} - e^{-2t} \end{pmatrix}.$$

$$(4) \qquad \Phi(t) = \begin{pmatrix} \frac{1}{2}(e^{-t} - e^{-3t}) & -\frac{1}{4}(e^{-t} + 2e^{3t}) \\ (-e^{-t} + e^{3t}) & \frac{1}{2}(e^{-t} + e^{3t}) \end{pmatrix}.$$

3.4. Solve the state space model:

$$\dot{x} = \begin{pmatrix} 1 & 1 \\ 0 & 0 \end{pmatrix} x + \begin{pmatrix} 0 \\ 1 \end{pmatrix} u$$
$$y = \begin{pmatrix} 1 & 0 \end{pmatrix} x.$$

The initial state is $x(0) = (1\ 1)^{\mathrm{T}}$. The input $u(t)$ is a unit step response.

3.5. Calculate $\Phi(t, 0)$ and $\Phi^{-1}(t, 0)$.

$$(1) \quad A = \begin{pmatrix} 2t & 0 \\ 0 & 1 \end{pmatrix}. \qquad (2) \quad A = \begin{pmatrix} 0 & e^{-t} \\ -e^{-t} & 0 \end{pmatrix}.$$

3.6. The discrete time system is listed below. Try to calculate $x(k)$.

$$\begin{bmatrix} x_1(k+1) \\ x_2(k+1) \end{bmatrix} = \begin{pmatrix} \frac{1}{3} & \frac{1}{9} \\ \frac{1}{9} & \frac{1}{3} \end{pmatrix} \begin{bmatrix} x_1(k) \\ x_2(k) \end{bmatrix} + \begin{pmatrix} 1 & 0 \\ 0 & 1 \end{pmatrix} \begin{bmatrix} u_1(k) \\ u_2(k) \end{bmatrix}$$

$$x_1(0) = -1, \quad x_2(0) = 4.$$

$u_1(k)$ is sampled from a ramp function t and $u_2(k)$ is sampled from e^{-t}.

4 Stability Analysis

4.1 Introduction

Stability is an important property for a system because, only when a system is stable can it finish the target task. In this chapter, a group of conceptions of stability in the sense of Lyapunov are given at the beginning, which are somewhat different from the definitions of stability given in classical control theory. Following that, the theorems to decide whether a system is stable or not are introduced.

4.2 Definition

The response of linear systems can always be decomposed as the zero state response and the zero input response. The stabilities of these two responses are commonly studied separately. The bounded input bounded output (BIBO) stability is for the zero state response, while marginal and asymptotic stabilities are for the zero input response.

Definition 4.1 (External Stability). An input $u(t)$ is said to be bounded if $u(t)$ does not grow to positive infinity or negative infinity. Equivalently, constants β_1 and β_2 exist, and

$$u(t) \leq \beta_1 < \infty \quad \text{holds for all} \quad t \geq 0 . \tag{4.1}$$

A system is said to be BIBO stable if every bounded input excites a bounded output. For example:

$$y(t) \leq \beta_2 < \infty \quad \text{holds for all} \quad t \geq 0 . \tag{4.2}$$

This stability is defined for the zero state response.

Conclusion 4.1 (BIBO Stability of Linear Time Variant System). *Consider a continuous linear time variant (LTV) system with p inputs, m outputs and zero initial condition. If we define $[t_0, \infty]$ as the time domain, the system is BIBO stable at time t_0 if, and only if, there exists a limited positive number β, which satisfies the following relationship:*

$$\int_{t_0}^{t} |h_{ij}(t, \tau)| \, d\tau \leq \beta < \infty , \tag{4.3}$$

where

$$h_{ij}(t, \tau) , \quad i = 1, 2, \ldots, m , \quad j = 1, 2, \ldots, p \tag{4.4}$$

are elements of the impulse response matrix $H(t, \tau)$ at any t ($t \in [t_0, \infty]$).

https://doi.org/10.1515/9783110574951-004

Proof. The proof is divided into two parts.

Step 1: The system is a SISO system. That is, it is $p = m = 1$.

First, if $h_{ij}(t, \tau)$ is absolutely integrable, then every bounded input excites a bounded output.

Suppose $u(t)$ is an arbitrary input with $u(t) \leq \beta_1 < \infty$ for all $t \geq 0$. This will lead to the output being bounded as follows:

$$|y(t)| = \left| \int_{t_0}^{t} h(t, \tau) u(\tau) d\tau \right| \leq \int_{t_0}^{t} |h(t, \tau)| \, |u(\tau)| \, d\tau$$

$$\leq \beta_1 \int_{t_0}^{t} |h(t, \tau)| \, d\tau \leq \beta_1 \beta = \beta_2 < \infty \tag{4.5}$$

Second, it can be seen that if $h_{ij}(t, \tau)$ is not absolutely integrable, the system is not BIBO stable.

If $h_{ij}(t, \tau)$ is not absolutely integrable, for any absolutely large N, there exists a $t_1 \in [t_0, \infty]$ such that:

$$\int_{t_0}^{t_1} |h(t_1, \tau)| \, d\tau \geq N \, .$$

Let us use the following example to demonstrate:

$$u(t) = \text{sgn } h(t_1, t) = \begin{cases} +1, & h(t_1, t) > 0 \\ 0, & h(t_1, t) = 0 \\ -1, & h(t_1, t) < 0 \, . \end{cases}$$

It is very clear that u is bounded. The output excited by the input is:

$$y(t_1) = \int_{t_0}^{t_1} h(t, \tau) u(\tau) d\tau = \int_{t_0}^{t_1} |h(t_1, \tau)| \, d\tau = \infty \, .$$

Because $y(t_1)$ can be absolutely large, we conclude that a similar bounded input can excite an unbounded output, which is contrary to the definition *external stability*. Therefore, the assumption does not hold, and we have:

$$\int_{t_0}^{t} |h_{ij}(t, \tau)| \, d\tau \leq \beta < \infty, \quad \forall t \in [t_0, \infty] \, .$$

Step 2: The system is a MIMO system. Note that any element $y_i(t)$ of output $y(t)$ is

$$|y_i(t)| = \left| \int_{t_0}^{t} [h_{i1}(t, \tau) u_1(\tau) + \cdots + h_{ip}(t, \tau) u_p(\tau)] \, d\tau \right| ,$$

$$\leq \left| \int_{t_0}^{t} h_{i1}(t, \tau) u_1(\tau) d\tau \right| + \cdots + \left| \int_{t_0}^{t} h_{ip}(t, \tau) u_p(\tau) d\tau \right| , \quad i = 1, 2, \ldots, m \, .$$

The sum of a finite number of bounded functions remains bounded. Therefore, we can have the conclusion based on the condition of SISO. Proof has been provided. □

Conclusion 4.2 (BIBO Stability of LTI System). *If we define the initial time as $t_0 = 0$ in a continuous LTI system with p inputs, m outputs and zero initial condition, the system is BIBO stable. However, this would only apply if, and only if, there exists a limited positive number of β, which satisfies the following relationship:*

$$\int_0^\infty |h_{ij}(t)|\, dt \le \beta < \infty ,$$

where $h_{ij}(t), i = 1, 2, \ldots, m, j = 1, 2, \ldots p$ are elements of the impulse response matrix $H(t)$.

Conclusion 4.3 (BIBO Stability of LTI System). *If we define the initial time as $t_0 = 0$ in a continuous time LTI system with p inputs, m outputs and zero initial condition, the system with proper rational transfer function matrix $G(s)$ is BIBO stable. However, this would only apply if, and only if, every pole of $G(s)$ has a negative real part or every pole of $G(s)$ lies in the left half s-plane.*

Proof. The characteristic polynomial of $G(s)$ is $\alpha_G(s)$. The pole of $G(s)$ is s_l ($l = 1, 2, \ldots, m$), which are the roots of $\alpha_G(s) = 0$. Therefore, any rational fraction of $G(s)$ is $g_{ij}(s)$ ($i = 1, 2, \ldots, q, j = 1, 2, \ldots p$). Its expansion contains the partial fractions:

$$\frac{\beta_l}{(s - s_l)^{\alpha_{lr}}} , \quad l = 1, 2, \ldots, m , \quad \alpha_{l_r} = 1, 2, \ldots, \sigma_l ,$$

where β_l is zero or a nonzero constant, and the pole s_l with multiplicity σ_l.

Thus, the inverse Laplace transform of $g_{ij}(s)$ is:

$$\rho_{lr} t^{\alpha_{lr}-1} e^{s_l t} , \quad l = 1, 2, \ldots, m .$$

If $\beta_l / (s - s_l)^{\alpha_{1_{lr}}} = \beta_l$, the corresponding inverse Laplace transform is the impulse function δ. Therefore, the element $h_{ij}(t)$ of the impulse response matrix $H(t)$, derived from the inverse Laplace transform of element transfer function $g_{ij}(s)$, is the sum of the finite terms as $\rho_{l,r} t^{\alpha_{lr}-1} e^{s_l t}$. It may contain the function δ. It is straightforward to verify that every $\rho_{l,r} t^{\alpha_{lr}-1} e^{s_l t}$ ($\forall i = 1, 2, \ldots, q, \forall j = 1, 2, \ldots, p$) is absolutely integrable if and only if the pole s_l ($l = 1, 2, \ldots, m$) has a negative real part, i.e., $h_{ij}(t)$ ($\forall i = 1, 2, \ldots, q, \forall j = 1, 2, \ldots, p$) is absolutely integrable.

Therefore, according to Conclusion 4.2, the system is BIBO stable. The BIBO stability is defined for the zero state response.

Now we study the stability of the zero input response, or the response of:

$$\dot{x}(t) = Ax(t) ,$$

which is excited by nonzero initial state x_0. Clearly, the solution is:

$$x(t) = e^{At} x_0 .$$ □

Definition 4.2 (Internal Stability). The zero input response of equation $\dot{x}(t) = Ax(t)$ is marginally stable, or stable in the sense of Lyapunov, if every finite internal state x_0 excites a bounded response. It is asymptotically stable if every finite initial state excites a bounded response, which also approaches 0 as $t \to \infty$.

Conclusion 4.4 (Internal Stability of LTV System). *The zero input response of equation $\dot{x}(t) = Ax(t)$ is internal stable or marginally stable if every finite internal state x_0 excites a bounded state transition matrix $\phi(t, t_0)$, which also approaches 0 as $t \to \infty$.*

Proof. If $x(t_0) = x_0$ at time t_0, the zero input response is:

$$x_{0u} = \phi(t, t_0)x_0, \forall t \in [t_0, \infty] \ .$$

It is straightforward to verify that x_{0u} is bounded, only in instances where $\phi(t, t_0)$ is bounded and $\lim_{t \to \infty} x_{0u}(t) = 0$, and when $\lim_{t \to \infty} \phi(t, t_0) = 0$. □

Conclusion 4.5 (Internal Stability of LTI System). *The zero input response of equation $\dot{x}(t) = Ax(t) + Bu$ with the initial state $x(0) = x_0$ is internal stable or marginal stable only in instances where $\lim_{t \to \infty} e^{At} = 0$.*

Proof. For a LTI system, the state transfer matrix $\phi(t) = e^{At}$ and e^{At} is bounded for any $t > 0$. Following this, we can obtain Conclusion 4.5 from Conclusion 4.4. □

Conclusion 4.6 (Internal Stability of LTI System). *The zero input response of equation $\dot{x}(t) = Ax(t) + Bu(t)$ with the initial state $x(0) = x_0$ is internal stable or marginal stable in instances where every eigenvalue $\lambda_i(A)$ $(i = 1, 2, \ldots, n)$ has a negative real part. In other words:*

$$\mathrm{Re}\,\{\lambda_i(A)\} < 0\,, \quad i = 1, 2, \ldots, n\,.$$

Conclusion 4.7 (The Relationship Between Internal Stability and External Stability). *Consider the continuous LTI system:*

$$\dot{x} = Ax + Bu\,, \quad x(0) = x_0\,, \quad t \geq 0\,,$$
$$y = Cx + Du\,,$$

where x is an n-dimensional state vector, u is a p-dimensional input vector, and y is a m-dimensional output vector. If the above system is internal stable or marginal stable, it must be BIBO stable or external stable.

Proof. For the above LTI system, from the analysis of the system dynamics we know that the impulse response matrix $H(t)$ is:

$$H(t) = Ce^{At}B + D\delta(t)\,.$$

From Conclusion 4.5 we know that if the system is internal stable, e^{At} is bounded and $\lim_{x \to \infty} e^{At} = 0$. With the above contents, we can get all elements of the impulse response matrix $H(t)$, where $h_{ij}(t)$ ($i = 1, 2, \ldots, m, j = 1, 2, \ldots, p$) satisfies the following relationship:

$$\int_0^\infty |h_{ij}(t)|\, dt \le \beta < \infty .$$

The system is BIBO stable according to Conclusion 4.2. □

Conclusion 4.8 (The Relationship Between External Stability and Internal Stability).
Consider the continuous LTI system:

$$\dot{x} = Ax + Bu , \quad x(0) = x_0 , \quad t \ge 0 ,$$
$$y = Cx + Du .$$

BIBO stability or external stability cannot guarantee internal stability or marginal stability.

Proof. When some poles and zeros are the same, the order of transfer function for a system is lower than that of state space description, i.e., the number of poles is less than the number of eigenvalues. The system is BIBO stable. In other words, every pole of $G(s)$ has a negative real part and cannot guarantee that the eigenvalues of the system have negative real parts. Therefore, BIBO stability cannot guarantee the internal stability of the system. □

Conclusion 4.9 (The Equivalence Between External Stability and Internal Stability).
Consider the continuous LTI system:

$$\dot{x} = Ax + Bu , \quad x(0) = x_0 , \quad t \ge 0 ,$$
$$y = Cx + Du .$$

Without the zero pole cancelation, the system is internal stable if, and only if, the system is external stable.

Proof. From Conclusion 4.7, we know that internal stability means external stability of the system. If the system has no zero pole cancelation, external stability means internal stability of the system according to the proof of Conclusion 4.8. Therefore, external stability is equivalent to internal stability of a system if the system has no zero pole cancelation. □

Definition 4.3 (Autonomous System). A dynamic system without external or input excitation is defined as an autonomous system.

Generally, the state equation of a continuous nonlinear time variant (NTV) autonomous system can be described as follows:

$$\dot{x} = f(x, t) , \quad x(t_0) = x_0 , \quad t \in [t_0, \infty] \tag{4.6}$$

where x is n-dimensional state vector, $f(x, t)$ is n-dimensional vector function. For a continuous nonlinear time invariant (NTI) system, state equation can be written as $\dot{x} = f(x)$.

For a continuous LTV system, the vector function $f(x, t)$ of equation (4.6) can be further described as a linear vector function of state x. The state equation of autonomous system can be rewritten as

$$\dot{x} = A(t)x, \quad x(t_0) = x_0, \quad t \in [t_0, \infty] \tag{4.7}$$

And the state equation of a continuous LTI autonomous can be written as $\dot{x} = Ax$.

Definition 4.4 (Equilibrium State). For a continuous NTV system, the equilibrium state of the autonomous system (4.7) is x_e, which satisfies the following equation:

$$\dot{x}_e = f(x, t) = 0, \quad \forall t \in [t_0, \infty] . \tag{4.8}$$

Below are some notes about the equilibrium state.
(a) Intuitive meaning of the equilibrium state: Equilibrium state x_e is a class of state which always satisfies $\dot{x}_e = 0$.
(b) The form of the equilibrium state: The equilibrium state x_e can be solved from equation (4.8). For a two-dimensional autonomous system, the form of x_e can be points or a line in the state space.
(c) Nonuniqueness: The equilibrium state x_e of an autonomous system is not always unique. For a continuous LTI system, the equilibrium state x_e is the solution of $Ax_e = 0$. If the matrix A is nonsingular, we have a unique solution of $x_e = 0$. If the matrix A is singular, the solution is not unique.
(d) Zero equilibrium state: For the autonomous systems (4.6) or (4.7), $x_e = 0$ must be an equilibrium state for the system.
(e) Isolated equilibrium states: The isolated equilibrium states are in the form of isolated equilibrium point in the state space. An important feature of the isolated equilibrium states is that they can be transferred to state space origin by moving coordinates.
(f) Agreement on the equilibrium states: In the direct method of Lyapunov, the stability analysis is mainly aimed at the equilibrium states. Therefore, we always set the state space origin as the equilibrium states, i.e., $x_e = 0$ in the following sections of the stability analysis.

Definition 4.5 (Disturbed Dynamics). The disturbed dynamics of a dynamic system is a class of state dynamics caused by the initial state x_0.

In nature, the disturbed dynamics is the state response of zero input. We call it disturbed dynamics because a nonzero initial state x_0 will be regarded as a state disturbance relative to the zero equilibrium state $x_e = 0$ in stability analysis.

Usually, for a more clear description of the relationship of time and causality in the disturbed dynamics, we further represent the disturbance dynamics in the following form:

$$x_{0u}(t) = \phi(t; x_0, t_0), \quad t \in [t_0, \infty] ,$$

where ϕ is a vector function. When $t = t_0$, the vector function of the disturbed dynamics satisfies $\phi(t_0; x_0, t_0) = x_0$.

In the sense of geometry, the disturbed dynamics $\phi(t; x_0, t_0)$ presents a trajectory from the initial state x_0 in the state space. We can constitute a trajectory cluster of disturbed dynamics $\phi(t; x_0, t_0)$ according to different initial states.

Definition 4.6 (the Stability in the Sense of Lyapunov). The isolated equilibrium state $x_e = 0$ of the autonomous system is considered to be stable in the sense of Lyapunov at the time instant t if, for any real number $\varepsilon > 0$, there exists a corresponding real number $\delta(\varepsilon, t_0) > 0$. When:

$$\|x_0 - x_e\| \le \delta(\varepsilon, t_0) , \tag{4.9}$$

the disturbed dynamics $\phi(t; x_0, t_0)$ from the initial x_0 satisfies the following inequality:

$$\|\phi(t; x_0, t_0) - x_e\| \le \varepsilon , \quad \forall t \ge t_0 . \tag{4.10}$$

Listed below are notes regarding the stability in the sense of Lyapunov.

(a) Geometric significance of stability
There is direct geometric significance about the stability in the sense of Lyapunov. Hence, inequality (4.10) can be considered as a supersphere in the state space, whose core is x_e and the radius is ε. Its field can be represented with $S(\varepsilon)$. Inequality (4.9) can be considered as a supersphere, whose core is x_e and the radius is $\delta(\varepsilon, t_0)$ in the state space. Its field can be represented with $S(\delta)$, which is a function of ε and t_0. The geometric explanation of stability in the sense of Lyapunov is that the dynamics trajectories $\phi(t; x_0, t_0)$ starting from any initial state within the field $S(\delta)$ will never exceed the boundary $H(\varepsilon)$ of the filed $S(\varepsilon)$, as shown in Figure 4.1.

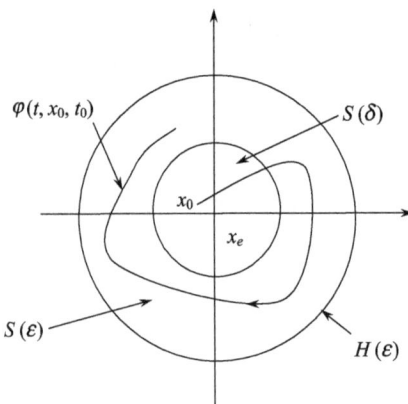

Fig. 4.1: The stability in the sense of Lyapunov.

(b) Uniform stability in the sense of Lyapunov
According to the definition of stability in the sense of Lyapunov, if a real number $\delta(\varepsilon) > 0$ exists, which is not related to the initial time t_0, i.e., when $\|x_0 - x_e\| \leq \delta(\varepsilon)$ holds, $\|\phi(t; x_0, t_0) - x_e\| \leq \varepsilon \forall t \geq t_0$ always holds. At this stage, we can call the equilibrium state x_e uniformly stable in the sense of Lyapunov. In general, for the time variant systems, uniform stability is of more practical significance than stability. Uniform stability means that, if the system is stable in the sense of Lyapunov at an initial time instant t_0, the system is stable in the sense of Lyapunov at all initial time t_0 within the definition interval of time.

(c) The stability properties of the time invariant system
For the time invariant system, whether it is a linear or nonlinear system or a continuous or time discrete system, stability in the sense of Lyapunov must be equivalent to uniform stability. In other words, if the equilibrium state x_e for a time invariant system is stable in the sense of Lyapunov, x_e must be uniformly stable in the sense of Lyapunov.

(d) The nature of stability in the sense of Lyapunov
The definition shows that the stability in the sense of Lyapunov can only guarantee the boundedness of the system's disturbed dynamics instead of the asymptotic characteristic relative to the equilibrium state. Therefore, stability in the sense of Lyapunov does not necessarily mean stability within industrial processes.

Definition 4.7 (the Asymptotic Stability). The isolated equilibrium state $x_e = 0$ of the autonomous system is considered to be asymptotic stable if the following conditions hold:
(i) If $x_e = 0$ is stable in the sense of Lyapunov at time t_0.
(ii) If, for a real number $\delta(\varepsilon, t_0) > 0$ and any real number $\mu > 0$, there exists a corresponding real number $T(\mu, \delta, t_0) > 0$. This would make the disturbed dynamics $\phi(t; x_0, t_0)$, starting from any initial state x_0 that satisfies the inequality (4.9), satisfy the following inequality:

$$\|\phi(t; x_0, t_0) - x_e\| \leq \mu, \quad \forall t \geq t_0 + T(\mu, \delta, t_0) . \tag{4.11}$$

We have provided the following points based on the definition of asymptotic stability.

(a) Geometric significance of asymptotic stability
Take a two-dimensional system for example. The geometric meaning of asymptotic stability is shown in Figures 4.2 and 4.3. The response from an initial state x_0 in the sphere $S(\delta)$ will not exceed the sphere $S(\varepsilon)$ (as shown in Figure 4.2) and converges to the sphere μ as time goes on (as shown in Figure 4.3).

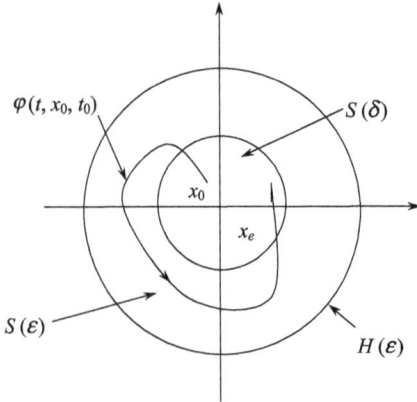

Fig. 4.2: The asymptotic stability.

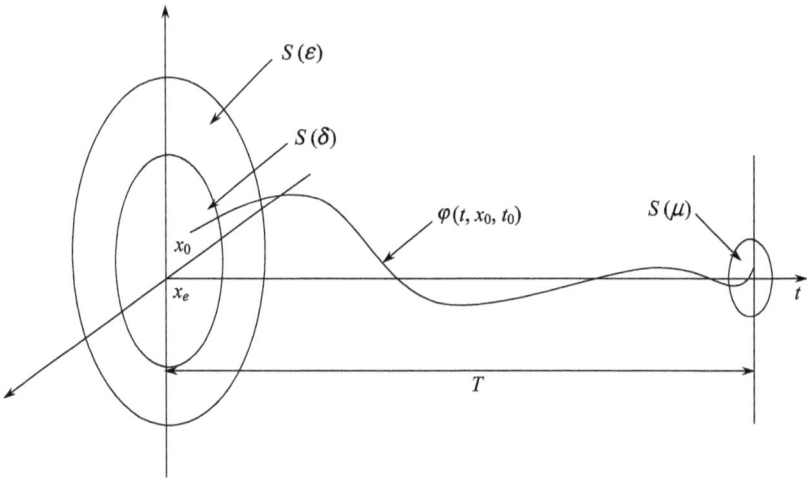

Fig. 4.3: The asymptotic stability.

(b) The equivalent definition of asymptotic stability

According to the definition of asymptotic stability, if we select $\mu \rightarrow 0$, then $T(\mu, \delta, t_0) \rightarrow \infty$. Therefore, the equivalent definition of the asymptotic stability can be introduced, which reflects the asymptotic characteristics of the stable process in a more intuitive form. The isolated equilibrium state $x_e = 0$ of an autonomous system (4.7) is asymptotic stable at time t_0 when two conditions hold. That is, when the disturbed dynamics $\phi(t; x_0, t_0)$ starting from any initial state $x_0 \in S(\delta)$ is bounded to any $t \in [t_0, \infty)$ relative to equilibrium state $x_e = 0$. And also when the disturbed dynamics relative to equilibrium state $x_e = 0$ meets the asymptotic characteristic, that is, $\lim_{t \to \infty} \phi(t; x_0, t_0) = 0, \forall x_0 \in S(\delta)$.

(c) Uniform asymptotic stability
In the definition of asymptotic stability, if $\delta(\varepsilon)$ has nothing to do with t_0, and the other conditions hold, the equilibrium state x_e is uniform asymptotic stable. Similarly, for time variant systems, the uniform asymptotic stability is more meaningful than the asymptotic stability.

(d) The properties of asymptotic stability for time invariant systems
For time invariant systems, whether the system is linear or nonlinear or time continuous or time discrete, the asymptotic stability is equivalent to the uniform asymptotic stability of the equilibrium state x_e. In other words, the asymptotic stability of the equilibrium state $x_e \Leftrightarrow$ is the uniform asymptotic stability of the equilibrium state x_e.

(e) Large scale and small scale asymptotic stability
Small scale asymptotic stability is also known as local asymptotic stability. The definition of local asymptotic stability is that:
There exists a supersphere $S(\delta)$ around $x_e = 0$, $\forall 0 \neq x_0 \in S(\delta)$,

$$x_e \text{ is asymptotically stable.} \tag{4.12}$$

Where $S(\delta)$ is the attraction domain, representing the property that all the states within $S(\delta)$ can be attracted to, the equilibrium state is x_e.
Large scale asymptotic stability is also known as global asymptotic stability. The definition of global asymptotic stability is:

$$\forall 0 \neq x_0 \in R^n, \quad x_e = 0. \quad \text{This is asymptotic stable.} \tag{4.13}$$

(f) The necessary condition of large scale asymptotic stability
From the definition of large scale asymptotic stability (4.13), the necessary condition for the equilibrium state $x_e = 0$ to be large scale asymptotic stable is that there are no other asymptotic stable equilibrium states in the state space R^n.

(g) The properties of asymptotic stability for linear systems
For linear systems, whether the system is time invariant or time variant, or if it is time continuous or time discrete, if the equilibrium state $x_e = 0$ is asymptotic stable, it is largescale asymptotic stable.

(h) The asymptotic stability in the sense of Lyapunov is $\Leftrightarrow$
The stability in the sense of engineering.

Definition 4.8 (the Instability). The isolated equilibrium state $x_e = 0$ of the autonomous system is considered to be unstable if, for $\varepsilon > 0$ (and if ε is big enough), a corresponding real number $\delta(\varepsilon, t_0) > 0$ does not exist. This makes the disturbed dynamics $\phi(t; x_0, t_0)$, starting from any initial state x_0 that satisfies the inequality $\|x_0 - x_e\| \leq \delta(\varepsilon, t_0)$, satisfy the following inequality:

$$\|\phi(t; x_0, t_0) - x_e\| \leq \varepsilon, \quad \forall t \geq t_0.$$

Take a two-dimensional system for example. The geometric meaning of instability is shown in Figure 4.4. If the equilibrium state $x_e = 0$ is unstable, no matter how large or small $S(\delta)$ is, nonzero point $x_0^* \in S(\delta)$ exists. This makes the disturbed dynamics trajectory starting from this point exceed the field $S(\delta)$. In essence, instability in the sense of Lyapunov is equivalent to divergent instability in the meaning of industrial process.

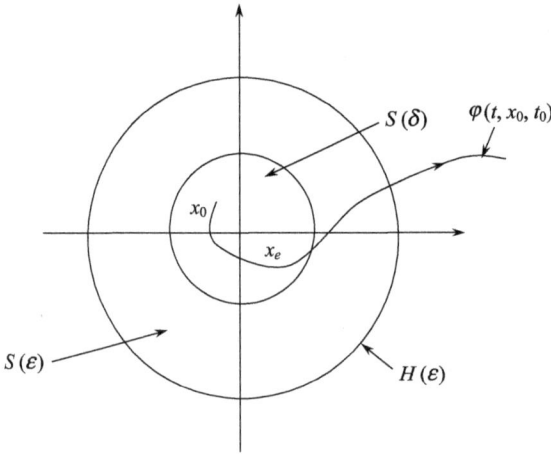

Fig. 4.4: Instability.

4.3 Stability Criteria

4.3.1 Lyapunov's Second Method

Lyapunov's second method proposes such a visual revelation in physics, similar to how the dynamic process of a system is accompanied by changes of energy. If the change rate of the system energy always remains negative, i.e., the energy decreases monotonously, the disturbed dynamics of the system will eventually return to the equilibrium state. Based on this fact, we present the following stability criteria.

Large Scale Asymptotic Stability Theorem
Consider a NTV autonomous system described by:

$$\dot{x} = f(x, t) , \quad t \in [t_0, \infty) , \tag{4.14}$$

where $x \in R^{n \times 1}$ and $f(0, t) = 0$ for all $t \in [t_0, \infty)$, which means the origin of the state space is an isolated equilibrium state.

Theorem 4.1. *The origin of* (4.14) *is large scale uniformly and asymptotically stable if a scalar function $V(x, t)$ exists that satisfies $V(0, t) = 0$ and has continuous first order partial derivatives for x and t, and qualifies the following terms for all the nonzero sates in state space R^n listed below:*

(i) *$V(x, t)$ is positive definite and bounded, i.e., two continuous nondecreasing scalar function $\alpha(\|x\|)$ and $\beta(\|x\|)$ ($\alpha(0) = 0$, $\beta(0) = 0$) exist, so that*

$$\beta(\|x\|) \geq V(x, t) \geq \alpha(\|x\|) > 0, \quad \text{for all} \quad x \neq 0 \quad \text{and} \quad t \in [t_0, \infty). \quad (4.15)$$

(ii) *The derivative of $V(x, t)$ on t: $\dot{V}(x, t)$ is negative definite and bounded, i.e., a continuous nondecreasing scalar function $y(\|x\|)$ ($y(0) = 0$) exists, so that:*

$$\dot{V}(x, t) \leq -y(\|x\|) < 0, \quad \text{for all} \quad x \neq 0 \quad \text{and} \quad t \in [t_0, \infty). \quad (4.16)$$

(iii) *$\alpha(\|x\|) \to \infty$ when $\|x\| \to \infty$; equivalently, $V(x, t) \to \infty$.*

Proof. Step 1: Prove that the origin equilibrium $x_e = 0$ is uniformly stable.

From the above term (i), we know that $\beta(\|x\|)$ is continuous nondecreasing and $\beta(0) = 0$. Thus, for any real number $\varepsilon > 0$, a real number $\delta(\varepsilon) > 0$ must exist so that $\beta(\delta) \leq \alpha(\varepsilon)$. Besides, as $\dot{V}(x, t)$ is negative definite, we have:

$$V(\phi(t; x_0, t_0), t) - V(x_0, t_0) = \int_{t_0}^{t} \dot{V}(\phi(\tau; x_0, t_0), \tau) d\tau \leq 0. \quad (4.17)$$

Then for any initial time t_0 and any nonzero initial state x_0 with $\|x_0\| \leq \delta(\varepsilon)$, we have

$$\alpha(\varepsilon) \geq \beta(\delta) \geq V(x_0, t_0) \geq V(\phi(t; x_0, t_0), t) \geq \alpha(\|\phi(t; x_0, t_0)\|), \quad \text{for any } t \in [t_0, \infty). \quad (4.18)$$

As $\alpha(\|x\|)$ is continuous nondecreasing and $\alpha(0) = 0$, we can use the above equation to deduce that, for any initial time t_0 and any nonzero initial state x_0 which has $\|x_0\| \leq \delta(\varepsilon)$, we have:

$$\|\phi(t; x_0, t_0)\| \leq \varepsilon, \quad \forall t \geq t_0. \quad (4.19)$$

Therefore, for any real number $\varepsilon > 0$, we can find a $\delta(\varepsilon) > 0$ ($\delta(\varepsilon)$, which is independent from the initial time t_0. This makes the response $\phi(t; x_0, t_0)$ excited by any initial time t_0 and any nonzero initial state x_0 with $\|x_0\| \leq \delta(\varepsilon)$ qualify equation (4.19). According to the definition, the origin equilibrium $x_e = 0$ is uniformly stable. Proof provided.

Step 2: Prove that for any initial time t_0, the dynamics $\phi(t; x_0, t_0)$ excited by any nonzero state x_0, which meets $\|x_0\| \leq \delta(\varepsilon)$, converges to the original equilibrium state $x_e = 0$.

First, for any real number $\mu > 0$ and the deduced real number $\delta(\varepsilon) > 0$, we can construct a real number $T(\mu, \delta) > 0$. Suppose the initial time t_0 is random and the nonzero x_0 satisfies $\|x_0\| \leq \delta(\varepsilon)$. Without loss of generality, we assume $0 < \mu \leq \|x_0\|$.

Then, as $V(x, t)$ is bounded, for the given $\mu > 0$, we can find a corresponding real number $v(\mu) > 0$, which makes $\beta(v) \leq \alpha(\mu)$. Besides, $\gamma(\|x\|)$ is continuous and nondecreasing. Suppose $\rho(\mu, \delta)$ is the minimum value of $\gamma(\|x\|)$ in the interval $v(\mu) \leq \|x\| \leq \varepsilon$. We can assume:

$$T(\mu, \delta) = \frac{\beta(\delta)}{\rho(\mu, \delta)} \, . \tag{4.20}$$

In accordance with this principle, for any given real number $\mu > 0$, we can construct a corresponding $T(\mu, \delta)$, which is independent of the initial time t_0.

Furthermore, for some time t_2 ($t_0 \leq t_2 \leq t_0 + T(\mu, \delta)$), we prove that $\phi(t_2; x_0, t_0) = v(\mu)$. Suppose that $t_1 = t_0 + T(\mu, \delta)$, and suppose that $\phi(t_2; x_0, t_0) > v(\mu)$ for any t in the interval $t_0 \leq t \leq t_1$. Then, using equation (4.20) and the negative definite property of $\dot{V}(x, t)$, we can deduce that:

$$\begin{aligned}
0 < \alpha(v) &\leq V(\phi(t_1; x_0, t_0), t_1) \leq V(x_0, t_1) \\
&\leq V(x_0, t_0) - (t_1 - t_0)\rho(\mu, \delta) \\
&\leq \beta(\delta) - T(\mu, \delta)\rho(\mu, \delta) \\
&= \beta(\delta) - \beta(\delta) = 0 \, .
\end{aligned} \tag{4.21}$$

Obviously, the above equation is a contradictory result. So the hypothesis does not hold, which means that a time t_2 in the time interval $t_0 \leq t \leq t_1$ must exist, which makes $\phi(t_2; x_0, t_0) = v(\mu)$.

Finally, we deduce that, for all $t \geq t_0 + T(\mu, \delta)$, we have $\|\phi(t; x_0, t_0)\| \leq \mu$. In this respect, considering $\phi(t_2; x_0, t_0) = v(\mu)$ and using the bound of $V(x, t)$ and the negative definite property of $\dot{V}(x, t)$, for all the $t \geq t_2$, we have:

$$\alpha\left(\|\phi(t; x_0, t_0)\|\right) \leq V(\phi(t; x_0, t_0), t) \leq V(\phi(t_2; x_0, t_0), t_2) \leq \beta(v) \leq \alpha(\mu) \, . \tag{4.22}$$

Thus, based on the fact that $\alpha(\|x\|)$ is a continuous nondecreasing function, we can deduce from equation (4.22) that for all $t \geq t_2$, we have:

$$\|\phi(t; x_0, t_0)\| \leq \mu \, . \tag{4.23}$$

Besides, from $t_0 + T(\mu, \delta) \geq t_2$ we can know that (4.23) holds for all t when $t \geq t_0 + T(\mu, \delta)$, and $T \to \infty$ when $\mu \to 0$.

As proven above, for any initial time instant t_0, the dynamics excited by any nonzero initial state x_0 with $\|x_0\| \leq \delta(\varepsilon)$ converges to the original equilibrium state $x_e = 0$ when $t \to \infty$.

Step 3: Prove that for any nonzero initial state x_0 in the state space R^n, its forced dynamics $\phi(t; x_0, t_0)$ is uniformly bounded.

As $\alpha(\|x\|) \to \infty$ when $\|x\| \to \infty$, there must exist a finite real number $\varepsilon(\delta) > 0$, which makes $\beta(\delta) < \alpha(\varepsilon)$ for any arbitrarily large real number $\delta > 0$. Using the bound of $V(x, t)$ and the negative definite property of $\dot{V}(x, t)$, we can know that for all $t \in [t_0, \infty)$ and any nonzero $x_0 \in R^n$, we have

$$\alpha(\varepsilon) > \beta(\delta) \geq V(x_0, t_0) \geq V(\phi(t; x_0, t_0), t) \geq \alpha(\|\phi(t; x_0, t_0)\|) \, . \tag{4.24}$$

Therefore, considering that $\alpha(\|x\|)$ is a continuous nondecreasing function, we have:

$$\|\phi(t; x_0, t_0)\| \le \varepsilon(\delta), \quad \forall t \ge t_0, \quad \forall x_0 \in R^n. \tag{4.25}$$

$\varepsilon(\delta)$ is independent of the initial time t_0. This indicates that for any nonzero initial state $x_0 \in R^n$, $\phi(t; x_0, t_0)$ is uniformly bounded. As such, all proof has been provided.

□

Notes on Theorem 4.1:

(a) Physical implication
For Theorem 4.1, in a physical sense, the positive definite bounded scalar function $V(x, t)$ is regarded as some kind of "generalized energy" and $\dot{V}(x, t)$ is regarded as the change rate of the generalized energy. This idea reflects an institutive fact that, if the energy of the system is limited and the change rate of the energy is negative definite, the system energy is bounded and eventually decreases to zero. Correspondingly, the dynamics of the system is bounded and eventually converges to the origin equilibrium.

(b) Lyapunov function
In Theorem 4.1, $V(x, t)$ is not equivalent to the energy. Furthermore, the meaning and form of $V(x, t)$ varies with the physical property. Thus, in the theory of system stability, $V(x, t)$ which qualifies the theorem is called the Lyapunov function. To judge the asymptotical stability of the system, we construct a Lyapunov function $V(x, t)$ for the system.

(c) The selection of Lyapunov function
For a comparatively simple system, we usually select a quadratic function of state x as the Lyapunov function. If the function does not satisfy the theorem, we can try to select a more complicated one. For a complex system, the construction of Lyapunov function is difficult. Therefore, frequently we select the Lyapunov function by method of trial and error.

(d) The sufficiency of the criterion
Theorem 4.1 is sufficient, but not a necessary condition, to judge the large scale uniform and asymptotic stability of the system (4.14). The limitation is that, if we cannot find the Lyapunov function $V(x, t)$ which meets the theorem, we cannot determine whether the system is stable or not.

(e) The principles in using the criterion
Considering the sufficient property of Theorem 4.1 in determining the stability, we first judge whether the system is large scale asymptotically stable or not. If the answer is no, then we judge the small scale asymptotical stability of the system. If the result is no again, we will judge whether the system is Lyapunov stable until the stability is determined. The above principle is helpful but does not always works.

Now, we discuss the continuous NTI system, for which the state equation is:

$$\dot{x} = f(x), \quad t \ge 0, \tag{4.26}$$

where $x \in R^{n \times n}$ and $f(0) = 0$ for all $t \in [0, \infty)$, i.e., the state space origin $x = 0$ is an isolated equilibrium state of the system.

We obtain the corresponding conclusion for the time invariant case directly from Theorem 4.1, since a time invariant system is a special case. Besides, we can see that the requirement is largely simplified in its form for time invariant cases.

Theorem 4.2. *For a continuous NTI autonomous system (4.26), if a scalar function $V(x)$ exists which has continuous first order partial derivatives for x, and qualifies the following terms for all the nonzero sates in state space R^n and $V(0) = 0$:*
(i) *$V(x)$ is positive definite*
(ii) *$\dot{V}(x) = dV(x)/dt$ is negative definite*
(iii) *$V(x) \to \infty$ when $\|x\| \to \infty$*
The original equilibrium state of (4.26) is large scale asymptotically stable.

Example 4.1. Consider a continuous NTI autonomous system:

$$\dot{x}_1 = x_2 - x_1(x_1^2 + x_2^2)$$
$$\dot{x}_2 = -x_1 - x_2(x_1^2 + x_2^2) .$$

Discuss the stability of the system.

Solution. Obviously, $[x_1, x_2]^T = [0, 0]^T$ is the equilibrium state.
First, we select a quadratic function of state x as the Lyapunov function $V(x)$:

$$V(x) = x_1^2 + x_2^2 .$$

It is obviously that $V(x)$ is positive definite.
Following that, by calculating $\dot{V}(x)$, we get:

$$\dot{V}(x) = \frac{\partial V(x)}{\partial x_1}\frac{dx_1}{dt} + \frac{\partial V(x)}{\partial x_2}\frac{dx_2}{dt} = \begin{bmatrix} \frac{\partial V(x)}{\partial x_1} & \frac{\partial V(x)}{\partial x_2} \end{bmatrix}\begin{bmatrix} \dot{x}_1 \\ \dot{x}_2 \end{bmatrix}$$

$$= \begin{bmatrix} 2x_1 & 2x_2 \end{bmatrix}\begin{bmatrix} x_2 - x_1(x_1^2 + x_2^2) \\ -x_1 - x_2(x_1^2 + x_2^2) \end{bmatrix} = -2(x_1^2 + x_2^2)^2 .$$

It is easy to see that $\dot{V}(x)$ is negative definite.
Lastly, when $\|x\| = \sqrt{x_1^2 + x_2^2} \to \infty$, we get:

$$V(x) = \|x\|^2 = (x_1^2 + x_2^2) \to \infty .$$

According to Theorem 4.2, the system original equilibrium state $x = 0$ is large scale asymptotically stable.

The major difficulty in constructing Lyapunov function is that the item $\dot{V}(x)$ should be negative definite. This is a quite conservative condition. Next, we will give a relaxed stable criterion for a continuous NTI system.

Theorem 4.3. *For a continuous NTI autonomous system (4.26), suppose a scalar function $V(x)$ exists. If $V(x)$ has continuous first order partial derivatives for x and qualifies the following terms for all the nonzero states in state space R^n and $V(0) = 0$:*
(i)   *$V(x)$ is positive definite*
(ii) *$\dot{V}(x) = dV(x)/dt$ is seminegative definite*
(iii) *$\dot{V}(\varphi(t, x_0, 0))$ is not identically equal to zero for any nonzero $x_0 \in R^n$*
(iv) *$V(x) \to \infty$ when $\|x\| \to \infty$*
the original equilibrium state of (4.26) is large scale asymptotically stable.

Example 4.2. Consider a continuous NTI autonomous system:

$$\dot{x}_1 = x_2$$
$$\dot{x}_2 = -x_1 - (1 + x_2)^2 x_2 .$$

Discuss the stability of the system.

Solution. Obviously, $[x_1, x_2]^T = [0, 0]^T$ is the only equilibrium state.
First, we select a quadratic function of state x as the Lyapunov function $V(x)$:

$$V(x) = x_1^2 + x_2^2 .$$

$V(x)$ is positive definite.
Second, by computation, we get:

$$\dot{V}(x) = \left[\frac{\partial V(x)}{\partial x_1} \quad \frac{\partial V(x)}{\partial x_2} \right] \begin{bmatrix} \dot{x}_1 \\ \dot{x}_2 \end{bmatrix}$$

$$= \begin{bmatrix} 2x_1 & 2x_2 \end{bmatrix} \begin{bmatrix} x_2 \\ -x_1 - (1 + x_2)^2 x_2 \end{bmatrix} = -2x_2^2 (1 + x_2)^2 .$$

We can see that there are two cases which make $\dot{V}(x) = 0$:
case 1: x_1 is arbitrary and $x_2 = 0$
case 2: x_1 is arbitrary and $x_2 = -1$

Except for these two cases, we have $\dot{V}(x) < 0$ when $x \neq 0$. Thus, $\dot{V}(x)$ is seminegative definite.
Now we check whether $\dot{V}(\varphi(t, x_0, 0))$ is identically equal to zero or not. The problem comes down to judging whether the above two cases are the disturbed response of the system.
For case 1,

$$\overline{\phi}(t; x_0, 0) = [x_1(t), 0]^T .$$

From $x_2(t) \equiv 0$, we can deduce $\dot{x}_2(t) = 0$. Substituting this into the system equation yields:

$$\dot{x}_1(t) = x_2(t) = 0$$
$$0 = \dot{x}_2(t) = -(1 + x_2(t))^2 x_2(t) - x_1(t) = -x_1(t) .$$

Therefore, $\overline{\phi}(t; x_0, 0) = [x_1(t), 0]^T$ is not the solution of the system disturbed dynamics except for the origin ($x_1 = 0, x_2 = 0$).

For case 2,

$$\overline{\phi}(t; x_0, 0) = [x_1(t), -1]^T .$$

From $x_2(t) = -1$, we can deduce that $\dot{x}_2(t) = 0$. Substituting this into the system equation yields:

$$\dot{x}_1(t) = x_2(t) = -1$$
$$0 = \dot{x}_2(t) = -(1 + x_2(t))^2 x_2(t) - x_1(t) = -x_1(t) .$$

Obviously, this is a contradictory result. Hence, $\overline{\phi}(t; x_0, 0) = [x_1(t), -1]^T$ is not the solution of the system. Therefore, item (iii) in Theorem 4.3 is satisfied.

Lastly, when $\|x\| = \sqrt{(x_1^2 + x_2^2)} \to \infty$, we get:

$$V(x) = \|x\|^2 = (x_1^2 + x_2^2) \to \infty .$$

According to Theorem 4.3, the original equilibrium state of the system $x = 0$ is large scale asymptotically stable. Besides, we can see that the Lyapunov function we chose for the system does not qualify for Theorem 4.2 but meets Theorem 4.3.

Small Scale Asymptotically Stable Theorem

In the application of the second Lyapunov method, when a system is not large scale asymptotically stable, we turn to judge the small scale asymptotically stable. This section presents some basic theorems about small scale asymptotically stable theorems in the application of the second Lyapunov method.

For continuous NTV systems, we have the following conclusion.

Theorem 4.4. *For a continuous NTV autonomous system (4.14), suppose a scalar function $V(x, t)$ ($V(0, t) = 0$) exists. If $V(x, t)$ has continuous first order partial derivatives for x and t and an attractive region called Ω around the state space origin, and meets the following requirements for all nonzero states $x \in \Omega$ and all $t \in [t_0, \infty)$:*
(i) $V(x, t)$ is positive definite and bounded
(ii) $\dot{V}(x, t) = dV(x, t)/dt$ is negative definite and bounded
then the original equilibrium state of the system $x = 0$ is uniformly and asymptotically stable in the Ω region.

For continuous NTI systems, we have the following two conclusions.

Theorem 4.5. *For a continuous NTI autonomous system (4.26), if a scalar function $V(x)$ ($V(0) = 0$) exists, which has continuous first order partial derivatives for x and t and an attractive region called Ω around the state space origin, and meets the following terms*

for all nonzero states x ∈ Ω and all t ∈ [t_0, ∞):
(i)   *V(x) is positive definite*
(ii) *$\dot{V}(x, t) = dV(x)/dt$ is negative definite*
then the original equilibrium state of the system x = 0 is asymptotically stable in the Ω region.

Theorem 4.6. *For a continuous NTI autonomous system (4.26), if a scalar function V(x) (V(0) = 0) exists, which has continuous first order partial derivatives for x and t and an attractive region called Ω around the state space origin, and meets the following criteria for all nonzero states x ∈ Ω and all t ∈ [t_0, ∞):*
(i)   *V(x) is positive definite*
(ii) *$\dot{V}(x, t) = dV(x)/dt$ is seminegative definite*
(iii) *$\dot{V}(\varphi(t, x_0, 0))$ is not identically equal to zero for any nonzero state x ∈ Ω*
then the original equilibrium state x = 0 is asymptotically stable in the region called Ω.

Theorem for Stability in the Sense of Lyapunov
Similar to when a system is not large scale asymptotically stable, when a system is not small scale asymptotically stable, we turn to judging the stability in the sense of Lyapunov. In this section, we will provide some rules to determine the stability in the sense of Lyapunov.

Theorem 4.7. *For a continuous NTV autonomous system (4.14), if a scalar function V(x, t) (V(0, t) = 0) exists, which has continuous first order partial derivatives for x and t and an attractive region called Ω around the state space origin, and meets the following terms for all the nonzero states x ∈ Ω and all t ∈ [t_0, ∞):*
(i)   *V(x, t) is positive definite and bounded*
(ii) *$\dot{V}(x, t) = dV(x, t)/dt$ is seminegative definite and bounded*
then the original equilibrium state of the system x = 0 is stable in the sense of Lyapunov in the Ω region.

For continuous NTV systems, we have the following conclusion.

Theorem 4.8. *For a continuous NTI autonomous system (4.26), if a scalar function V(x) exists, V(0) = 0, V(x) has continuous first order partial derivatives for x and t and an attractive region called Ω around the state space origin which qualifies the following terms for all nonzero states x ∈ Ω and all t ∈ [t_0, ∞):*
(i)   *V(x) is positive definite*
(ii) *$\dot{V}(x) = dV(x)/dt$ is seminegative definite*
then the original equilibrium state of the system x = 0 is stable in the sense of Lyapunov in the region called Ω.

Theorem for Instability
For a continuous NTV system, the criterion for instability is presented as follows.

Theorem 4.9. *For a continuous NTV autonomous system (4.14), if a scalar function*
$V(x, t)$ *($V(0, t) = 0$) exists, which has continuous first order partial derivatives for x*
and t and an attractive region called Ω around the state space origin, and meets the
following terms for all the nonzero sates $x \in \Omega$ and all $t \in [t_0, \infty)$:
(i) $V(x, t)$ is positive definite and bounded
(ii) $\dot{V}(x, t) = dV(x, t)/dt$ is positive definite and bounded
then the original equilibrium state of the system $x = 0$ is unstable.

For continuous NTV systems, we have the following criteria.

Theorem 4.10. *For a continuous NTI autonomous system (4.26), if a scalar function*
$V(x)$ *($V(0) = 0$) exists, which has continuous first order partial derivatives for x and*
t and an attractive region called Ω around the state space origin, and meets the follow-
ing terms for all nonzero states $x \in \Omega$ and all $t \in [t_0, \infty)$:
(i) $V(x)$ is positive definite
(ii) $\dot{V}(x) = dV(x)/dt$ is positive definite
then the original equilibrium state of the system $x = 0$ is unstable.

Note: From the above two conclusions, we can see that the system is unstable when
$V(x, t)$ or $V(x)$ have the same sign with $\dot{V}(x, t)$ or $\dot{V}(x)$. Theoretically, the disturbed
dynamics trajectories of the system will diverge to infinity.

4.3.2 State Dynamics Stability Criteria for Continuous Linear Systems

This section discusses the stability for continuous linear systems. Based on the concepts and results of the second Lyapunov method, similar to the LTI and LTV system,
we will discuss the stability of the disturbed dynamics first. Then some stable criteria
will be presented.

Stability Criteria for LTI Systems
Consider a continuous LTI system. The autonomous state equation is:

$$\dot{x} = Ax, \quad x(0) = x_0, \quad t \geq 0, \tag{4.27}$$

where $x \in R^n$, and the origin of the state space $x = 0$ is an equilibrium state of the
system.

Next, we present the stability criteria for LTI systems based on eigenvalues.

Theorem 4.11. *For a continuous LTI system (4.27), the original equilibrium state $x = 0$ is stable in the sense of Lyapunov if, and only if, all the eigenvalues of matrix A have nonpositive real parts, i.e., zero or negative real parts, and the eigenvalue, whose real part is zero, is distinct.*

Proof. The proof is divided into two steps.

Step 1: Prove that the system is stable if $\|e^{At}\| \leq \beta < \infty$. From the autonomous dynamics equation of the LTI system, we can obtain the disturbed dynamics of states:

$$\phi(t; x_0, 0) = x_{0u}(t) = e^{At} x_0 . \tag{4.28}$$

The equilibrium state is $x_e = 0$. We notice that $x_e = e^{At} x_e$, thus we further deduce that the disturbed dynamics relative to equilibrium state $x_e = 0$ is:

$$\phi(t; x_0, 0) - x_e = e^{At}(x_e - x_0), \quad \forall t \geq 0 . \tag{4.29}$$

This indicates that, specifically if $\|e^{At}\| \leq \beta < \infty$, for any real number ε a real number $\delta(\varepsilon) = \varepsilon/\beta$ exists, which is independent of the initial time and makes the disturbed dynamics from any nonzero initial state $\|x_0 - x_e\| \leq \delta(\varepsilon)$ ($x_0 \in R^n$) qualify for the following inequality:

$$\left\| \phi(t; x_0, 0) - x_e \right\| \leq \left\| e^{At} \right\| \cdot \|x_0 - x_e\| \leq \beta \cdot \frac{\varepsilon}{\beta} = \varepsilon, \quad \forall t \geq 0 . \tag{4.30}$$

As defined, the system is stable in the sense of Lyapunov. Proof has been successfully provided.

Step 2: Prove the conclusion of Theorem 4.11. Introduce the linear nonsingular transformation $\hat{x} = Q^{-1} x$ to ensure $\widehat{A} = Q^{-1} A Q$ as the Jordan Canonical:

$$\left\| e^{\widehat{A}t} \right\| \leq \left\| Q^{-1} \right\| \left\| e^{At} \right\| \|Q\|, \quad \left\| e^{At} \right\| \leq \|Q\| \left\| e^{\widehat{A}t} \right\| \left\| Q^{-1} \right\| . \tag{4.31}$$

This indicates that, the bound of $\|e^{At}\|$ is equivalent to the bound of $\|e^{\widehat{A}t}\|$. From the Jordan Canonical, we know that the element of $e^{\widehat{A}t}$ is the combination of the following items:

$$t^{\beta_i - 1} e^{\alpha_i t + j\omega_i t}, \quad \lambda_i(\widehat{A}) = \lambda_i(A) = \alpha_i + j\omega_i, \quad i = 1, 2, \ldots, \mu, \quad \beta_i = 1, 2, \ldots, \sigma_i, \tag{4.32}$$

where $\lambda(\cdot)$ is the eigenvalue of the corresponding matrix, and σ_i means that λ_i is a σ_i duplicate eigenvalue. When $\alpha_i < 0$, the corresponding items are bounded in the interval $[0, \infty)$ for any limited positive integral β_i. When $\alpha_i = 0$, the corresponding items are bounded in the interval $[0, \infty)$ only for $\beta_i = 1$. Furthermore, the bound of the elements of $e^{\widehat{A}t}$ means the bound of $\|e^{\widehat{A}t}\|$. This indicates that, $\|e^{\widehat{A}t}\|$, i.e., $\|e^{At}\|$ is bounded only if all the eigenvalues of matrix A have zero or negative real parts and the eigenvalues whose parts are zero are distinct. Using the proposition given in the first part, we can prove that the above condition is the necessary, and sufficient, condition for the stability in the sense of Lyapunov. Proof has been successfully provided. □

Theorem 4.12. *For a continuous LTI system (4.27), the original equilibrium state $x = 0$ is asymptotically stable, only if all the eigenvalues of matrix A have negative real parts.*

Proof. From Theorem 4.11, the equilibrium state $x = 0$ is stable in the sense of Lyapunov only if all the eigenvalues of matrix A have zero or negative real parts, and the eigenvalues whose real parts are zero are distinct. Furthermore, from equations (4.28), (4.31) and (4.32), we know that:

$$\lim_{t \to \infty} \phi(t; x_0, 0) = \lim_{t \to \infty} e^{At} x_0 = 0 \,,$$

$$\Leftrightarrow \quad \lim_{t \to \infty} \left\| e^{At} \right\| = 0$$

$$\Leftrightarrow \quad \lim_{t \to \infty} t^{\beta_i - 1} e^{\alpha_i t + j \omega_i t} = 0, i = 1, 2, \ldots, \mu \,, \quad \beta_i = 1, 2, \ldots, \sigma_i \,.$$

$$\Leftrightarrow \quad \text{The eigenvalues of } A \text{ all have negative real parts.}$$

As defined, the system is asymptotically stable. Proof has been successfully provided.

□

Note: We can see that the asymptotic stability equals to the internal stability illustrated beforehand. Furthermore, based on the second Lyapunov method, we can provide the Lyapunov stability criteria for LTI systems.

Theorem 4.13. *For an n-dimensional continuous LTI system (4.27), the original equilibrium state $x_e = 0$ is asymptotically stable only if, for any given $n \times n$ dimensional positive definite symmetry matrix Q, the Lyapunov equation*

$$A^T P + PA = -Q$$

has a unique $n \times n$ dimensional positive definite symmetry matrix solution P.

Proof. First we prove the sufficiency. Given $n \times n$ positive definite matrix P, we want to prove the asymptotic stability of $x_e = 0$. For this, we select the Lyapunov function $V(x) = x^T P x$. As $P = P^T > 0$. $V(x)$ is positive definite. Furthermore, we have:

$$V(x) = \dot{x}^T P x + x^T P \dot{x} = (Ax)^T P x + x^T P(Ax)$$
$$= x^T (A^T P + PA) x = -x^T Q x \,. \tag{4.33}$$

Moreover, from $Q = Q^T > 0$ we know that $\dot{V}(x)$ is negative. According to large scale asymptotically stable theorems, $x_e = 0$ is asymptotically stable. Sufficiency has been proven.

Then we prove the necessity. Given the asymptotic stability of $x_e = 0$, we want to prove that matrix P is positive definite. For this, we use the matrix equation:

$$\dot{X} = A^T X + XA \,, \quad X(0) = Q \,, \quad t \geq 0 \,. \tag{4.34}$$

The matrix X is:

$$X(t) = e^{A^T t} Q e^{At} \,, \quad t \geq 0 \,. \tag{4.35}$$

The integration of (4.34) from $t = 0$ to $t \to \infty$ is:

$$X(\infty) - X(0) = A^{\mathrm{T}} \left(\int_0^\infty X(t)dt \right) + \left(\int_0^\infty X(t)dt \right) A . \tag{4.36}$$

As the system is asymptotically stable, e.g., $e^{At} \to 0$ when $t \to \infty$, from (4.35) we have $X(\infty) = 0$. Considering $X(0) = Q$, let $P = \int_0^\infty X(t)dt$, so (4.36) can be further expressed as:

$$A^{\mathrm{T}} P + P A = -Q . \tag{4.37}$$

Hence, $P = \int_0^\infty X(t)dt$ is the solution of the Lyapunov equation. The fact is, $X(t)$ is unique and $X(\infty) = 0$, $P = \int_0^\infty X(t)dt$ is unique. While:

$$P^{\mathrm{T}} = \int_0^\infty \left[e^{A^{\mathrm{T}}t} Q e^{At} \right]^{\mathrm{T}} dt = \int_0^\infty e^{A^{\mathrm{T}}t} Q e^{At} dt = P . \tag{4.38}$$

So $P = \int_0^\infty X(t)dt$ is symmetry. Again, for any nonzero $x_0 \in R^n$, we have:

$$x_0^{\mathrm{T}} P x_0 = \int_0^\infty (e^{At}x_0)^{\mathrm{T}} Q (e^{At}x_0) dt , \tag{4.39}$$

where the positive definite matrix $Q = N^{\mathrm{T}}N$ is nonsingular. From equation (4.39), we can further deduce:

$$x_0^{\mathrm{T}} P x_0 = \int_0^\infty (e^{At}x_0)^{\mathrm{T}} N^{\mathrm{T}} N (e^{At}x_0) dt$$

$$= \int_0^\infty \left\| N e^{At} x_0 \right\|^2 dt > 0 . \tag{4.40}$$

Therefore, P is unique and positive definite. Necessity has been proven and proof, overall, has been provided. □

Example 4.3. Consider the stability of the following continuous LTI system:

$$\dot{x} = \begin{bmatrix} -1 & 1 \\ 2 & -3 \end{bmatrix} x .$$

Solution. For simplicity, we select $Q = I_2$. Furthermore, from the Lyapunov equation:

$$A^{\mathrm{T}} P + P A = \begin{bmatrix} -1 & 2 \\ 1 & -3 \end{bmatrix} \begin{bmatrix} p_1 & p_3 \\ p_3 & p_2 \end{bmatrix} + \begin{bmatrix} p_1 & p_3 \\ p_3 & p_2 \end{bmatrix} \begin{bmatrix} -1 & 1 \\ 2 & -3 \end{bmatrix} = \begin{bmatrix} -1 & 0 \\ 0 & -1 \end{bmatrix} = -Q ,$$

we can deduce:

$$-2p_1 + 0p_2 + 4p_3 = -1$$
$$0p_1 - 6p_2 + 2p_3 = -1$$
$$p_1 + 2p_2 - 4p_3 = 0 .$$

Using algebraic equation solving methods, we get:

$$
\begin{bmatrix} p_1 \\ p_2 \\ p_3 \end{bmatrix} = \begin{bmatrix} -2 & 0 & 4 \\ 0 & -6 & 2 \\ 1 & 2 & -4 \end{bmatrix}^{-1} \begin{bmatrix} -1 \\ -1 \\ 0 \end{bmatrix} = \begin{bmatrix} -\frac{5}{4} & -\frac{1}{2} & -\frac{3}{2} \\ -\frac{1}{8} & -\frac{1}{4} & -\frac{1}{4} \\ -\frac{3}{8} & -\frac{1}{4} & -\frac{3}{4} \end{bmatrix} \begin{bmatrix} -1 \\ -1 \\ 0 \end{bmatrix} = \begin{bmatrix} \frac{7}{4} \\ \frac{3}{8} \\ \frac{5}{8} \end{bmatrix}.
$$

Therefore, solution of the Lyapunov equation is:

$$
P = \begin{bmatrix} \frac{7}{4} & \frac{5}{8} \\ \frac{5}{8} & \frac{3}{8} \end{bmatrix} > 0.
$$

P is positive definite, so the system is asymptotically stable.

MATLAB can be adopted for the solution of the above question.

```
>> A={[-1 1;2 -3]};
>> Q={[1 0;0 1]};
>> P=lyap(A',Q)
P =
    1.7500 0.6250
    0.6250 0.3750
```

The function "posdef" can be used to judge whether a matrix is positive definite or not.

The format of the MATLAB function is:

```
[key,sdet]=posdef(P)
```

The codes are shown as follows:

```
function [key,sdet]=posdef(P)
[nr,nc]=size(P);
sdet=[];
for i=1:nr
    sdet=[sdet,det(P(1:i,1:i))];
end
key=1;
if any(sdet<=0)
    key=0;
end
```

The running result of posdef(P) is:

```
key =
    1
sdet =
    1.7500    0.2656
```

If key=1, the result represents that P is positive definite. Otherwise P is negative definite. Sdet is the determinant of every matrix in the upper left corner.

Theorem 4.14. *For an n-dimensional continuous LTI system (4.27) and any given real number $\sigma \geq 0$, suppose the eigenvalues of matrix A are $\lambda_i(A)$, $i = 1, 2, \ldots, n$. Then all the eigenvalues located in the left half plane of the straight line $-\sigma + j\omega$ on s plane, i.e.,*

$$\operatorname{Re} \lambda_i(A) < -\sigma , \quad i = 1, 2, \ldots, n$$

if and only if for any given $n \times n$ dimensional positive definite symmetry matrix Q, the expanded Lyapunov function

$$2\sigma P + A^T P + PA = -Q \tag{4.41}$$

has unique positive definite solution matrix P.

Proof. Suppose $\widetilde{A} = A + \sigma I$, then:

$$\det(\tilde{s}I - \widetilde{A}) = \det(\tilde{s}I - A - \sigma I) = \det\left[(\tilde{s} - \sigma)I - A\right]$$
$$= \det(sI - A) , \quad \tilde{s} = s + \sigma . \tag{4.42}$$

From this, we know that:

$$\lambda_i(\widetilde{A}) = \lambda_i(A) + \sigma , \quad i = 1, 2, \ldots, n . \tag{4.43}$$

From Theorem 4.13, we know that all the eigenvalues of matrix $\widetilde{A}$ have negative real parts only if, for any positive definite symmetry matrix Q, the following Lyapunov function has unique positive definite solution matrix P:

$$\widetilde{A}^T P + P\widetilde{A} = -Q \tag{4.44}$$

Hence, substituting $\widetilde{A} = A + \sigma I$ into (4.44), we can deduce (4.41). While, from (4.43), we have the following equivalent relationship:

$$\operatorname{Re} \lambda_i(\widetilde{A}) < 0 \Leftrightarrow \operatorname{Re} \lambda_i(A) < -\sigma , \quad i = 1, 2, \ldots, n .$$

Therefore, if (4.41) has unique positive definite solution matrix P, $\operatorname{Re} \lambda_i(A) < -\sigma$, $i = 1, 2, \ldots, n$. Proof has been provided. $\qquad\square$

Stability Criteria for LTV Systems

Now we turn to discuss the continuous LTV systems. The autonomous state equation is:

$$\dot{x} = A(t)x , \quad x(t_0) = x_0 , \quad t \in [t_0, \infty] , \quad t_0 \in [t_0, \infty) , \tag{4.45}$$

where $x \in R^{n \times n}$. $A(t)$ qualifies the condition, which guarantees the existence and uniqueness of the solution and $x_e = 0$ is an equilibrium state of the system. Usually, there is a nonzero equilibrium state x_e besides $x_e = 0$.

For LTV systems, we can adopt two ways to judge the stability of equilibrium states; the method based on state transfer matrix and the method based on Lyapunov criteria. Next, we will introduce these two methods.

Theorem 4.15. *For a continuous LTV system (4.45), $\phi(t, t_0)$ is the state transfer matrix of the system. The original equilibrium state $x_e = 0$ is stable in the sense of Lyapunov at time t_0, only if a real number $\beta(t_0) > 0$ exists, which makes the following equation valid:*

$$\|\phi(t, t_0)\| \le \beta(t_0) < \infty , \quad \forall t \ge t_0 . \tag{4.46}$$

Furthermore, if there exists independent real numbers $\beta > 0$ for all t_0, the original equilibrium state $x_e = 0$ is stable in the sense of Lyapunov.

Theorem 4.16. *For a continuous LTV system (4.45), $\phi(t, t_0)$ is the state transfer matrix of the system. Then the original equilibrium state $x_e = 0$ is asymptotically stable at time t_0, only if a real number $\beta(t_0) > 0$ exists, which qualifies the following two items:*

$$\|\phi(t, t_0)\| \le \beta(t_0) < \infty, \forall t \ge t_0$$
$$\lim_{t \to \infty} \|\phi(t, t_0)\| = 0 . \tag{4.47}$$

Furthermore, the original equilibrium state $x_e = 0$ is uniformly and asymptotically stable, only if independent real numbers $\beta_1 > 0$ and $\beta_2 > 0$ exist for all $t_0 \in [0, \infty]$ that qualify the following equation:

$$\|\phi(t, t_0)\| \le \beta_1 e^{-\beta_2(t-t_0)} . \tag{4.48}$$

Proof. First we prove the sufficiency. Given equation (4.47), we need to prove that $x_e = 0$ is uniformly and asymptotically stable. From (4.48), and using the disturbed dynamics equation, we have:

$$\|\phi(t; x_0, t_0)\| = \|\phi(t, t_0)x_0\| \le \|\phi(t, t_0)\| \|x_0\| \le \beta_1 \|x_0\| e^{-\beta_2(t-t_0)} . \tag{4.49}$$

This indicates that the disturbed dynamics $\phi(t; x_0, t_0)$ is bounded for all $t \ge t_0$. For all $t_0 \in [0, \infty)$ we have $\|\phi(t; x_0, t_0)\| \to 0$ when $t \to \infty$. Thus, $x_e = 0$ is uniformly and asymptotically stable. Sufficiency has been proven.

Now we will prove the necessity. Given that $x_e = 0$ is uniformly and asymptotically stable, we need to prove equation (4.47). As $x_e = 0$ is uniformly and asymptotically stable, $x_e = 0$ is stable in the sense of Lyapunov, i.e., there exists a real number $\beta_3 > 0$ which satisfies:

$$\|\phi(t, t_0)\| \le \beta_3 , \quad \forall t_0 \in [0, \infty] , \quad \forall t \ge t_0 . \tag{4.50}$$

Furthermore, for a fixed real number $\delta > 0$ and any given real $\mu > 0$, there exists a real number $T > 0$ that satisfies the following equation for all initial states x_0 and all $t_0 \in [0, \infty)$:

$$\|\phi(t_0 + T; x_0, t_0)\| = \|\phi(t_0 + T; t_0, x_0)\| \le \mu . \tag{4.51}$$

Randomly select a x_0 to satisfy:

$$\|x_0\| = \delta \quad \text{and} \quad \|\phi(t_0 + T, t_0)x_0\| = \|\phi(t_0 + T, t_0)\| \cdot \|x_0\| . \tag{4.52}$$

Then, by selecting $\mu = \delta/2$ from equations (4.51) and (4.52), we can further deduce that:

$$\|\phi(t_0 + T, t_0)\| \leq \frac{1}{2}, \quad \forall t_0 \in [0, \infty) . \tag{4.53}$$

Hence, using equations (4.50) and (4.53) we can get:

$$\|\phi(t, t_0)\| \leq \beta_3 , \quad \forall t \in [t_0, t_0 + T) ,$$

$$\|\phi(t, t_0)\| = \|\phi(t, t_0 + T)\phi(t_0 + T, t_0)\| ,$$

$$\leq \|\phi(t, t_0 + T)\| \|\phi(t_0 + T, t_0)\| \leq \frac{\beta_3}{2} , \quad \forall t \in [t_0 + T, t_0 + 2T) ,$$

$$\|\phi(t, t_0)\| \leq \|\phi(t, t_0 + 2T)\| \|\phi(t_0 + 2T, t_0 + T)\| \|\phi(t_0 + T), t_0\| \leq \frac{\beta_3}{2^2} ,$$

$$\forall t \in [t_0 + 2T, t_0 + 3T) ,$$

$$\vdots$$

$$\|\phi(t, t_0)\| \leq \frac{\beta_3}{2^m} , \quad \forall t \in [t_0 + mT, t_0 + (m + 1)T) .$$

Again we construct an exponential function $\beta_1 e^{-\beta_2(t-t_0)}$, which makes the following equation valid:

$$\left[\beta_1 e^{-\beta_2(t-t_0)}\right]_{t=t_0+mT} = \frac{\beta_3}{2^{m-1}} , \quad m = 1, 2, \quad \dots . \tag{4.54}$$

Furthermore, we can get:

$$\beta_1(e^{-\beta_2 T})^m = 2\beta_3 \left(\frac{1}{2}\right)^m . \tag{4.55}$$

We can see that, by selecting $\beta_1 = 2\beta_3$ and an adequate β_2, we can make $e^{-\beta_2 T} = 1/2$ hold. Thus, we proved that real numbers $\beta_1 > 0$ and $\beta_2 > 0$ exist to validate equation (4.48). Necessity has been proven and proof, overall, has been provided. □

Theorem 4.17. *For a continuous LTV system (4.45), suppose $x_e = 0$ is the unique equilibrium state of the system. The elements of $n \times n$ dimensional matrix $A(t)$ are segmented continuous uniform and bounded real function, and the original equilibrium state $x_e = 0$ is uniformly and asymptotically stable if two real numbers, $\beta_1 > 0$ and $\beta_2 > 0$, exist when $0 < \beta_1 I \leq Q(t) \leq \beta_2 I$ holds. The $n \times n$ solution matrix $P(t)$ of the Lyapunov equation:*

$$-\dot{P}(t) = P(t)A(t) + A^\mathsf{T}(t)P(t) + Q(t) , \quad \forall t \geq t_0 , \tag{4.56}$$

is real symmetry, uniformly bounded and uniformly positive definite. Equivalently, two real numbers exist, $\alpha_1 > 0$ and $\alpha_2 > 0$, making $0 < \alpha_1 I \leq P(t) \leq \alpha_2 I, \forall t \geq t_0$.

4.3.3 State Dynamics Stability Criteria for Discrete Systems

Lyapunov Stability Theorem for Discrete NTI Systems
Consider a discrete NTI system. The autonomous equation is:

$$x(k + 1) = f(x(k)), \quad x(0) = x_0, \quad k = 0, 1, 2, \ldots, \tag{4.57}$$

where $x \in R^{n \times n}$, $f(0) = 0$, i.e., the origin of the state space $x = 0$ is an equilibrium state.

Next, we will present some Lyapunov stability theorems for discrete NTI systems.

Theorem 4.18. *For a discrete NTI system (4.57), if a scalar function $V(x(k))$ exists for discrete state $x(k)$ which meets the following items for any $x(k) \in R^n$:*
(i) $V(x(k))$ is positive definite
(ii) if $\Delta V(x(k)) = V(x(k + 1)) - V(x(k))$, $\Delta V(x(k))$ is negative
(iii) $V(x(k)) \to \infty$ when $x(k) \to \infty$
then the original equilibrium state $x = 0$ is large scale asymptotically stable.

Note: The conservative property of (ii) may result in the failure of judgment for many systems. Thus, we can release this condition as follows.

Theorem 4.19. *For a discrete NTI system (4.57), if a scalar function $V(x(k))$ exists for discrete state $x(k)$, which meets the following items for any $x(k) \in R^n$:*
(i) $V(x(k))$ is positive definite
(ii) if $\Delta V(x(k)) = V(x(k + 1)) - V(x(k))$, $\Delta V(x(k))$ is seminegative
(iii) $\Delta V(x(k))$ is not identically zero for any free dynamics started from any nonzero initial state $x(0) \in R^n$, i.e., equation (4.57)
(iv) $V(x(k)) \to \infty$ when $x(k) \to \infty$
then the original equilibrium state $x = 0$ is large scale asymptotically stable.

Based on the above stability theorems, we can easily deduce a more intuitive and convenient stability criterion for discrete systems.

Theorem 4.20. *For a discrete NTI system (4.57), suppose $f(0)=0$, $x = 0$ is an equilibrium state of the system. If $f(x(k))$ is convergent, i.e., for $x(k) \neq 0$, we have:*

$$\|f(x(k))\| < \|x(k)\| . \tag{4.58}$$

Then the original equilibrium state $x = 0$ is large scale asymptotically stable.

Proof. For a given discrete system, we select the Lyapunov function:

$$V(x(k)) = \|x(k)\| .$$

Obviously, $V(x(k))$ is positive definite. Furthermore, we can deduce that:

$$\Delta V(x(k)) = V(x(k + 1)) - V(x(k)) = \|x(k + 1)\| - \|x(k)\|$$
$$= \|f(x(k))\| - \|x(k)\| .$$

From equation (4.58), we can see that $\Delta V(x(k))$ is negative, and $V(x(k)) \to \infty$ when $x(k) \to \infty$. According to Theorem 4.18, the original equilibrium state $x = 0$ is large scale asymptotically stable. Proof has been provided. □

Stability Criteria for Discrete LTI Systems

Consider a discrete NTI system. The autonomous equation is:

$$x(k + 1) = Gx(k) , \quad x(0) = x_0 , \quad k = 0, 1, 2, \ldots , \tag{4.59}$$

where $x \in R^{n \times n}$, and the solution state x_e of $Gx_e = 0$ is an equilibrium state. If the matrix G is singular, there are nonzero equilibrium states besides $x_e = 0$. While, if the matrix G is nonsingular, there is only one equilibrium state $x_e = 0$.

Next, we will give the corresponding equilibrium state stability criteria for LTI systems.

Theorem 4.21. *For a discrete LTI autonomous system (4.59), the original equilibrium state $x_e = 0$ is stable in the sense of Lyapunov, only if all the amplitudes of the eigenvalues of G: $\lambda_i(G)$ ($i = 1, 2, \ldots, n$) are equal to or less than 1, and the eigenvalue whose amplitude is 1 is the single root of the polynomial of G.*

Theorem 4.22. *For a discrete LTI autonomous system (4.59), the origin equilibriums state $x_e = 0$ is asymptotically stable, only if all the amplitudes of the eigenvalues of G: $\lambda_i(G)$ ($i = 1, 2, \ldots, n$) are less than 1.*

Theorem 4.23. *For an n-dimensional discrete LTI autonomous system (4.59), the origin equilibriums state $x_e = 0$ is asymptotically stable. That is, all the amplitudes of the eigenvalues of G: $\lambda_i(G)$ ($i = 1, 2, \ldots, n$) are less than 1, but only if any given $n \times n$ dimensional positive definite symmetry matrix Q the discrete Lyapunov function*

$$G^T PG - P = -Q \tag{4.60}$$

has unique $n \times n$ dimensional positive definite symmetry solution matrix P.

Theorem 4.24. *For an n-dimensional discrete LTI autonomous system (4.59), the original equilibrium state $x_e = 0$ is exponentially stable with the index of $\sigma > 0$. That is, the eigenvalues of G satisfy*

$$|\lambda_i(G)| < \sigma , \quad 0 \le \sigma \le 1 , \quad i = 1, 2, \ldots, n , \tag{4.61}$$

but only if, for any given $n \times n$ dimensional positive definite symmetry matrix Q, the expanded discrete Lyapunov function

$$(1/\sigma)^2 G^T PG - P = -Q \tag{4.62}$$

has unique $n \times n$ dimensional positive definite symmetry solution matrix P.

Example 4.4. The state equation of a discrete linear system is:

$$x(k + 1) = \begin{bmatrix} \lambda_1 & 0 \\ 0 & \lambda_2 \end{bmatrix} x(k)$$

Try to determine the condition for the asymptotic stability of the equilibrium state.

Solution. According to $G^T P G - P = -I$, we have:

$$\begin{bmatrix} \lambda_1 & 0 \\ 0 & \lambda_2 \end{bmatrix} \begin{bmatrix} p_{11} & p_{12} \\ p_{12} & p_{22} \end{bmatrix} \begin{bmatrix} \lambda_1 & 0 \\ 0 & \lambda_2 \end{bmatrix} - \begin{bmatrix} p_{11} & p_{12} \\ p_{12} & p_{22} \end{bmatrix} = \begin{bmatrix} -1 & 0 \\ 0 & -1 \end{bmatrix}.$$

Then:

$$p_{11}(1 - \lambda_1^2) = 1 ; \quad p_{12}(1 - \lambda_1 \lambda_2) = 0 ; \quad p_{22}(1 - \lambda_2^2) = 1 .$$

So

$$P = \begin{bmatrix} \frac{1}{1-\lambda_1^2} & 0 \\ 0 & \frac{1}{1-\lambda_2^2} \end{bmatrix}.$$

The condition for asymptotic stability of the equilibrium state is:

$$|\lambda_1| < 1 \quad \text{and} \quad |\lambda_2| < 1 .$$

MATLAB can be adopted for the solution of the above question.

`P=dlyap(G',Q);`

If P is positive definite, the system is asymptotically stable.

4.4 Summary

Stability is very important for a system. In this chapter, the definitions of stability in the sense of Lyapunov were given for equilibrium state. Different stable criteria were listed and proven for different kinds of systems and examples were selected to show how to use the criteria.

Exercise

4.1. Determine whether the following functions are positive definite or not.

(1) $V(x) = 2x_1^2 + 3x_2^2 + x_3^2 - 2x_1 x_2 + 2x_1 x_3$

(2) $V(x) = \frac{1}{2} \left[(x_1 + x_2)^2 + 2x_1^2 + x_2^2 \right]$

(3) $V(x) = x_1^2 + x_3^2 - 2x_1 x_2 + x_2 x_3$

(4) $V(x) = x_1^2 + 3x_2^2 + 11x_3^2 - 2x_1 x_2 + 4x_2 x_3 + 2x_1 x_3$

4.2. Given a continuous time NTI system, try to analyze the stability of its equilibrium state:
$$\dot{x}_1 = x_2$$
$$\dot{x}_2 = -x_1^2 x_2 - x_1 .$$

4.3. Consider a continuous time NTI system:

$$\dot{x}_1 = x_2$$
$$\dot{x}_2 = -x_1 - x_2(1 + x_2)^2 .$$

Try to determine the stability of the original equilibrium state $x_e = 0$.

4.4. Consider a continuous time LTI system:

$$\dot{x} = \begin{bmatrix} 0 & 1 \\ -1 & -1 \end{bmatrix} x .$$

Try to determine the stability of its equilibrium state.

4.5. Consider a continuous time NTI system:

$$\dot{x}_1 = x_2$$
$$\dot{x}_2 = -(1 - |x_1|)x_2 - x_1 .$$

Try to analyze the stability of its equilibrium state.

4.6. Given the state space equation:

$$\dot{x} = \begin{bmatrix} 1 & 1 \\ -1 & 1 \end{bmatrix} x ,$$

try to determine the stability of the original equilibrium state $x_e = 0$.

4.7. Given the state space equation:

$$\dot{x} = \begin{bmatrix} 0 & 1 \\ -2 & -3 \end{bmatrix} x ,$$

try to determine the stability of the equilibrium point.

4.8. Consider a continuous time linear time varying system:

$$\dot{x} = \begin{bmatrix} 0 & 1 \\ -\frac{1}{t+1} & -10 \end{bmatrix} x , \quad t \geq 0 .$$

Try to determine whether the original equilibrium state $x_e = 0$ is large scale asymptotically stable. (Hint: Suppose $V(x, t) = \frac{1}{2}[x_1^2 + (1 + t)x_2^2]$.)

4.9. Try to analyze the BIBO stability and the asymptotical stability of the system equilibrium state $x_e = 0$ of the following two systems:

$$(1) \quad \dot{x} = \begin{bmatrix} 0 & 6 \\ 1 & -1 \end{bmatrix} x + \begin{bmatrix} -2 \\ 1 \end{bmatrix} u, \qquad (2) \quad \dot{x} = \begin{bmatrix} 0 & 1 & 0 \\ 0 & 0 & 1 \\ 250 & 0 & -5 \end{bmatrix} x + \begin{bmatrix} 0 \\ 0 \\ 10 \end{bmatrix} u$$

$$y = \begin{bmatrix} 0 & 1 \end{bmatrix} x, \qquad\qquad\qquad y = \begin{bmatrix} -25 & 5 & 0 \end{bmatrix} x,$$

4.10. Consider a linear discrete time system:

$$x(k+1) = \begin{bmatrix} \lambda_1 & 0 \\ 0 & \lambda_2 \end{bmatrix} x(k).$$

Try to determine the asymptotically stable condition for the equilibrium state.

4.11. Given a discrete time LTI system:

$$x(k+1) = \begin{bmatrix} 1 & 4 & 0 \\ -3 & -2 & -3 \\ 2 & 0 & 0 \end{bmatrix} x(k).$$

Use two methods to determine whether the system is asymptotically stable.

5 Controllability and Observability

5.1 Introduction

This chapter introduces the concepts of controllability and observability. Controllability deals with whether or not the state of a state space equation can be controlled from the input, and observability deals with whether or not the initial state can be observed from the output. These concepts can be illustrated using the network shown in Figure 5.1. In Figure 5.1 (a), the network has two state variables. Suppose that x_i is the voltage across the capacitor with capacitance C_i, for $i = 1, 2$. The input u is a voltage source. From the network, it can be seen that, when $C_1 = C_2$, $R_1 = R_2$, we always have $x_1 = x_2$. The input u cannot change x_1 and x_2 to any value, i.e., the system is uncontrollable. In Figure 5.1 (b), when the initial value of x_1 and x_2 have $x_1(t) = x_2(t)$, the output cannot reflect the value of $x_1(t)$ and $x_2(t)$, so the system is unobservable.

These concepts are essential in describing the internal structure of linear systems. They are also needed in studying control and filtering problems. In this chapter, we will discuss continuous time LTI state equations.

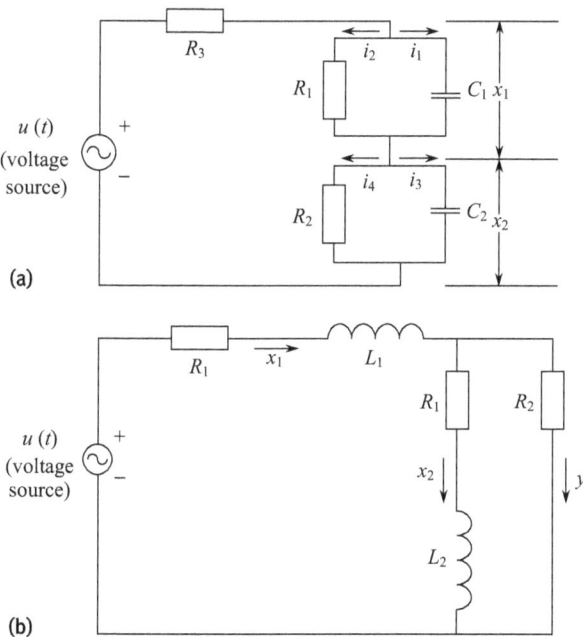

(a)

(b)

Fig. 5.1: (a) Network; (b) Network.

https://doi.org/10.1515/9783110574951-005

5.2 Definition

5.2.1 Controllability

Consider the n-dimensional p-input state equation:

$$\dot{x} = Ax + Bu \,, \tag{5.1}$$

where A and B are, respectively, $n \times n$ and $n \times p$ are real constant matrices. Because the output does not play any role in controllability, we will disregard the output equation in this study.

Definition 5.1. The state equation (5.1) or the pair (A, B) is said to be controllable if, for any initial state $x(0) = x_0$ and any final state x_1, there exists an input that transfers x_0 to x_1 in a finite time. Otherwise (5.1) or (A, B) is said to be uncontrollable.

This definition requires only that the input be capable of moving any state in the state space to any other state in finite time (what trajectory the state should take is not specified). Furthermore, there is no constraint imposed on the input; its magnitude can be as large as desired.

Consider the n-dimensional p-input q-output state equation:

$$\begin{aligned} \dot{x} &= Ax + Bu \\ y &= Cx + Du \,, \end{aligned} \tag{5.2}$$

where A, B, C and D are, respectively, $n \times n, n \times p, q \times n$ and $q \times p$ constant matrices.

Example 5.1. Consider the network shown in Figure 5.2 (a). Its state variable x is the voltage across the capacitor. If $x(0) = 0$, then $x(t) = 0$ for all $t \geq 0$, no matter what

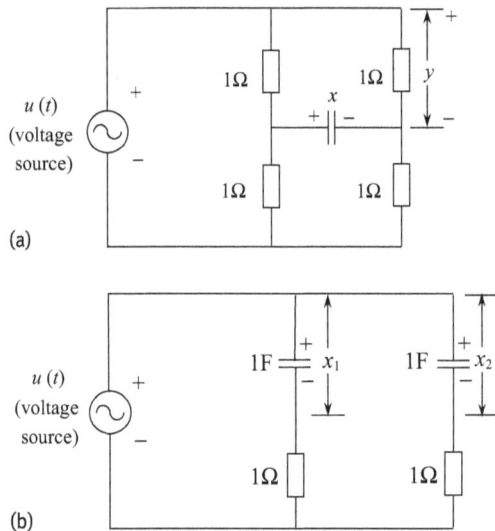

(a)

(b)

Fig. 5.2: Uncontrollable networks.

input is applied. This is due to the symmetry of the network. Furthermore, the input has no effect on the voltage across the capacitor. Thus the system or, more precisely, the state equation that describes the system., is not controllable.

Next, we consider the network shown in Figure 5.2 (b). It has two state variables: x_1 and x_2. The input can transfer x_1 or x_2 to any values, but it cannot transfer both x_1 and x_2 to any values. For example, if $x_1(0) = x_2(0) = 0$, then no matter what input is applied, $x_1(t)$ always equals $x_2(t)$ for all $t \geq 0$. Thus, the equation that describes the network is not controllable.

5.2.2 Observability

Definition 5.2. The state equation (5.2) is said to be observable if, for any unknown initial state $x(0)$, there exists a finite $t_1 > 0$ such that the knowledge of the input u and the output y over $[0, t_1]$ suffices to uniquely determine the initial state $x(0)$. Otherwise, the equation is said to be unobservable.

Example 5.2. Consider the network shown in Figure 5.3. If the input is zero, no matter what the initial voltage across the capacitor is, the output is identically zero because of the symmetry of the four resistors. We know the input and output (both are identically zero), but we cannot uniquely determine the initial state. Thus the network or, more precisely, the state equation that describes the network is not observable.

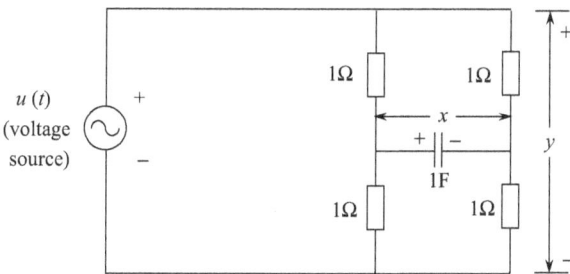

Fig. 5.3: Unobservable network.

The response of (5.2) excited by the initial state $x(0)$ and the input $u(t)$ is:

$$y(t) = Ce^{At}x(0) + C \int_0^t e^{A(t-\tau)}Bu(\tau)d\tau + Du(t) . \tag{5.3}$$

In the study of observability, the output y and the input u are assumed to be known; the initial state $x(0)$ is the only unknown one. Thus, we can write (5.3) as:

$$Ce^{At}x(0) = \bar{y}(t) , \tag{5.4}$$

where

$$\bar{y}(t) = y(t) - C \int_0^t e^{A(t-\tau)} Bu(\tau) d\tau - Du(t)$$

is a known function. Thus, the observability problem reduces to solving $x(0)$ from (5.4). If $u \equiv 0$, then $\bar{y}(t)$ reduces to the zero input response $Ce^{At}x(0)$. Therefore, Definition 5.2 can be modified as follows: Equation (5.2) is observable if, and only if, the initial state $x(0)$ can be determined uniquely from its zero input response over a finite time interval.

5.3 Criteria

5.3.1 Controllable Criteria

Consider the continuous LTI system – the state equation is expressed as:

$$\dot{x} = Ax + Bu, \quad x(0) = x_0, \quad t \geq 0, \tag{5.5}$$

where $x \in R^n$; $u \in R^r$; $A_{n \times n}$, $B_{n \times r}$.

Theorem 5.1 (Controllability Gram Matrix Criteria). *System (5.5) is controllable only if the $n \times n$ matrix*

$$W_c[0, t_1] \triangleq \int_0^{t_1} e^{-At} BB^T e^{-A^T t} dt \tag{5.6}$$

is nonsingular for any $t_1 > 0$.

Proof. First we show that, if $W_c[0, t_1]$ is nonsingular, then (5.5) is controllable. For any nonzero state x_0, the response of (5.5) at time t_1 is derived as:

$$x(t_1) = e^{At_1} x_0 + \int_0^{t_1} e^{A(t_1-t)} Bu(t) dt$$

$$= e^{At_1} x_0 - \left[e^{At_1} \int_0^{t_1} e^{-At} BB^T e^{A^T t} dt \right] W_c^{-1}[0, t_1] x_0$$

$$= e^{At_1} x_0 - e^{At_1} W_c[0, t_1] W_c^{-1}[0, t_1] x_0$$

$$= e^{At_1} x_0 - e^{At_1} x_0 = 0, \quad \forall x_0 \in R^n. \tag{5.7}$$

This shows that all nonzero states in R^n are controllable. As defined, the system is completely controllable. We show the converse by contradiction. Suppose $W_c[0, t_1]$ is singular, and that there exists a nonzero state $\bar{x}_0$, such that $\bar{x}_0^T W_c[0, t_1] x_0 = 0$. This would leave us with:

$$0 = \overline{x}_0^T W_c[0, t_1]x_0 = \int_0^{t_1} \overline{x}_0^T e^{-At} BB^T e^{-A^Tt} \overline{x}_0 dt$$

$$= \int_0^{t_1} \left[B^T e^{-A^Tt} \overline{x}_0 \right]^T \left[B^T e^{-A^Tt} \overline{x}_0 \right] dt$$

$$= \int_0^{t_1} \left\| B^T e^{-A^Tt} \overline{x}_0 \right\|^2 dt , \tag{5.8}$$

which implies that:

$$B^T e^{-A^Tt} \overline{x}_0 = 0 , \quad \forall t \in [0, t_1] . \tag{5.9}$$

If (5.5) is controllable, an input exists that ensures:

$$0 = x(t_1) = e^{At_1} \overline{x}_0 + \int_0^{t_1} e^{At_1} e^{-At} Bu(t)dt , \tag{5.10}$$

thus

$$\overline{x}_0 = - \int_0^{t_1} e^{-At} Bu(t)dt$$

$$\| \overline{x}_0 \|^2 = \overline{x}_0^T \overline{x}_0 = \left[- \int_0^{t_1} e^{-At} Bu(t)dt \right]^T \overline{x}_0 = - \int_0^{t_1} u^T(t)[B^T e^{-At} \overline{x}_0]dt . \tag{5.11}$$

From equation (5.9), (5.11) can be derived as:

$$\| \overline{x}_0 \|^2 = 0 \quad \Rightarrow \quad \overline{x}_0 = 0 , \tag{5.12}$$

which contradicts $\overline{x}_0 \neq 0$. Thus $W_c[0, t_1]$ is nonsingular. Proof has been provided. □

Theorem 5.2 (Controllability Rank Criteria). *The system (5.5) is controllable only if the $n \times np$ controllability matrix*

$$Q_c = \begin{bmatrix} B & AB & A^2B & \cdots & A^{n-1}B \end{bmatrix} \tag{5.13}$$

has rank n (full row rank).

Proof. First we show that, if rank $Q_c = n$, then (5.5) is controllable.

Suppose the system is not completely controllable. Using the Gram Matrix Criteria, we can get that the Gram matrix:

$$W_c[0, t_1] \triangleq \int_0^{t_1} e^{-At} BB^T e^{-A^Tt} dt , \quad \forall t_1 > 0 \tag{5.14}$$

is singular, which means that a nonzero state α exists, such that:

$$0 = \alpha^T W_c[0, t_1]\alpha = \int_0^{t_1} \alpha^T e^{-At} BB^T e^{-A^T t} dt$$

$$= \int_0^{t_1} \left[\alpha^T e^{-At} B\right]\left[\alpha^T e^{-At} B\right]^T dt. \qquad (5.15)$$

Therefore, we have:

$$\alpha^T e^{-At} B = 0, \quad \forall t \in [0, t_1]. \qquad (5.16)$$

Compute the $n - 1$ order derivative of the above equation and let $t = 0$, and we have:

$$\alpha^T B = 0, \quad \alpha^T AB = 0, \quad \alpha^T A^2 B = 0, \quad \dots, \quad \alpha^T A^{n-1} B = 0,$$

which equals:

$$\alpha^T \begin{bmatrix} B & AB & A^2 B & \dots & A^{n-1} B \end{bmatrix} = \alpha^T Q_c = 0. \qquad (5.17)$$

Because $\alpha \neq 0$, we know that all the rows of Q_c are linearly independent and equivalently, rank $Q_c < n$, which contradicts the hypothesis that rank $Q_c = n$. So the system is controllable. We can also prove the converse by contradiction.

Suppose that rank $Q_c < n$, and that there exists a nonzero state α, such that:

$$\alpha^T Q_c = \alpha^T \begin{bmatrix} B & AB & A^2 B & \dots & A^{n-1} B \end{bmatrix} = 0,$$

which implies:

$$\alpha^T A^i B = 0, \quad i = 0, 1, \dots, n - 1. \qquad (5.18)$$

Thus, for any $t_1 > 0$, we have:

$$\pm \frac{A^i t^i}{i!} B = 0, \quad \forall t \in [0, t_1], \quad i = 0, 1, 2, \dots,$$

or:

$$0 = \alpha^T \left[I - At + \frac{1}{2!}A^2 t^2 - \frac{1}{3!}A^3 t^3 + \dots\right] B = \alpha^T e^{-At} B, \quad \forall t \in [0, t_1]. \qquad (5.19)$$

Hence, we can get:

$$0 = \alpha^T \int_0^{t_1} e^{-At} BB^T e^{-A^T t} \alpha dt = \alpha^T W_c[0, t_1]\alpha, \qquad (5.20)$$

which indicates that the Gram matrix $W_c[0, t_1]$ is singular, and that the system is not totally controllable. This contradicts the hypothesis that the system is controllable. Proof has been provided. $\square$

Theorem 5.3 (Controllability PBH Criteria). *The system (5.5) is controllable only if:*

$$\text{rank}[sI - A, B] = n, \quad \forall s \in \ell, \tag{5.21}$$

or if:

$$\text{rank}[\lambda_i I - A, B] = n, \quad i = 1, 2, \ldots, n, \tag{5.22}$$

where ℓ is plural field and λ_i ($i = 1, 2, \ldots, n$) is the eigenvalue.

Proof. First we prove that, if (5.5) is controllable, then equation (5.21) and (5.22) are correct.

Suppose for a certain eigenvalue λ_i, $\text{rank}[\lambda_i I - A, B] < n$ exists, which implies that all the rows of Q_c are linearly independent. Hence, there must exist a nonzero n-dimensional constant vector α, such that:

$$\alpha^T [\lambda_i I - A, B] = 0. \tag{5.23}$$

That is, $\alpha^T A = \lambda_i \alpha^T$, $\alpha^T B = 0$.

Furthermore, $\alpha^T B = 0$, $\alpha^T AB = \lambda_i \alpha^T B = 0, \ldots, \alpha^T A^{n-1} B = 0$, which equals:

$$\alpha^T \begin{bmatrix} B & AB & \ldots & A^{n-1} B \end{bmatrix} = \alpha^T Q_c = 0. \tag{5.24}$$

Because $\alpha \neq 0$, we have rank $Q_c < n$.

From the rank criteria, we know that the system is completely controllable and, therefore, the hypothesis is not supported. Besides, for all s in plural field ℓ except the eigenvalues λ_i, we have $\text{rank}[sI - A, B] = n$, so (5.22) equals (5.21).

Conversely, we suppose the system is not completely controllable, and that there must exist a linear nonsingular transformation which transforms (A, B) into the following form:

$$\bar{A} = PAP^{-1} = \begin{bmatrix} \bar{A}_c & \bar{A}_{12} \\ 0 & \bar{A}_{\bar{c}} \end{bmatrix}$$

$$\bar{B} = PB = \begin{bmatrix} \bar{B}_c \\ 0 \end{bmatrix}, \tag{5.25}$$

where ($\bar{A}_c \in R^{n \times n}$, $\bar{B}_c \in R^{n \times p}$) and ($\bar{A}_{\bar{c}} \in R^{(n-h) \times (n-h)}$, $\bar{B}_c \in R^{(n-h) \times p}$), respectively, denote the controllable part and uncontrollable part after being decomposed. And the eigenvalue of A and $\bar{A}_{\bar{c}}$ have the following relationship:

$$\lambda_i = \text{an eigenvalue of } \bar{A}_{\bar{c}} = \text{an eigenvalue of } A$$

Define $\bar{q}_{\bar{c}} \in \ell^{1 \times (n-h)}$ is one characteristic vector of λ_i

We can construct a nonzero n-dimensional row vector:

$$q^T = \begin{bmatrix} 0, \bar{q}_{\bar{c}}^T \end{bmatrix} P \cdot P^{-1} \begin{bmatrix} \bar{B}_c \\ 0 \end{bmatrix} = 0$$

$$q^T A = \begin{bmatrix} 0, \bar{q}_{\bar{c}}^T \end{bmatrix} P \cdot P^{-1} \begin{bmatrix} \bar{A}_c & \bar{A}_{12} \\ 0 & \bar{A}_{\bar{c}} \end{bmatrix} P$$

$$= \begin{bmatrix} 0, \bar{q}_{\bar{c}}^T \bar{A}_{\bar{c}} \end{bmatrix} P = \begin{bmatrix} 0, \lambda_i \bar{q}_{\bar{c}}^T \end{bmatrix} P = \lambda_i \begin{bmatrix} 0, \bar{q}_{\bar{c}}^T \end{bmatrix} P = \lambda_i q^T.$$

This shows that an n-dimensional row vector $q^T = 0$ exists, such that:

$$q^T [\lambda_i I - A, B] = 0 . \tag{5.26}$$

Equivalently, a $\lambda_i \in \ell$ exists, such that:

$$\text{rank} [\lambda_i I - A, B] < n . \tag{5.27}$$

Obviously, this contradicts "rank$[sI - A, B] = n, \forall s \in \ell$", thus the hypothesis is not established and (5.5) is controllable. Proof has been provided. □

Theorem 5.4 (Controllability and the Jordan Canonical Form Criteria I). *Consider system* (5.5). *Suppose that the n eigenvalues $\lambda_1, \lambda_2, \ldots, \lambda_n$ are pairwise differently, and that the system is controllable, but only if the Jordan canonical of (5.5) is:*

$$\dot{\overline{x}} = \begin{bmatrix} \lambda_1 & & & \\ & \lambda_2 & & \\ & & \ddots & \\ & & & \lambda_n \end{bmatrix} \overline{x} + \overline{B}u , \tag{5.28}$$

where $\overline{B}$ does not contain zero row vector. This means each row vector of $\overline{B}$ satisfies:

$$\overline{b}_i \neq 0 , \quad i = 1, 2, \ldots, n . \tag{5.29}$$

Proof. For the Jordan canonical (5.28), we construct the PBH Criteria matrix:

$$\left[sI - \overline{A}, \overline{B} \right] = \begin{bmatrix} s - \lambda_i & & & \overline{b}_1 \\ & s - \lambda_2 & & \overline{b}_2 \\ & & \ddots & \vdots \\ & & & s - \lambda_n & \overline{b}_n \end{bmatrix} . \tag{5.30}$$

From the unit structure which makes up the matrix, we have $s = \lambda_i, i \in [1, 2, \ldots, n]$, rank$[sI - A, B] = n$, but only if "$b_i \neq 0, \forall i \in [1, 2, \ldots, n]$". Proof has been provided. □

Theorem 5.5 (Controllability and the Jordan Canonical Form Criteria II). *Consider system* (5.5). *Suppose the n eigenvalues are $\lambda_1 (\sigma_1$ layers, α_1 layers), $\lambda_2 (\sigma_2$ layers, α_2 layers), $\ldots, \lambda_l (\sigma_l$ layers, α_l layers), and $\sigma_1 + \sigma_2 + \cdots + \sigma_l = n, \lambda_i \neq \lambda_j, \forall i \neq j, i, j = 1, 2, \ldots, l$. Assume that the following Jordan canonical form is derived from linear nonsingular transformation of state equation (5.5):*

$$\dot{\hat{x}} = \widehat{A}\hat{x} + \widehat{B}u , \tag{5.31}$$

where:

$$\underset{n \times n}{\widehat{A}} = \begin{bmatrix} J_1 & & & \\ & J_2 & & \\ & & \ddots & \\ & & & J_l \end{bmatrix} , \qquad \underset{n \times p}{\widehat{B}} = \begin{bmatrix} \widehat{B}_1 \\ \widehat{B}_2 \\ \vdots \\ \widehat{B}_l \end{bmatrix} , \tag{5.32}$$

$$
\underset{\sigma_i \times \sigma_i}{J_i} = i \begin{bmatrix} J_{i1} & & & \\ & J_{i2} & & \\ & & \ddots & \\ & & & J_{i\alpha_i} \end{bmatrix}, \qquad \underset{\sigma_i \times p}{\widehat{B}_i} = \begin{bmatrix} \widehat{B}_{i1} \\ \widehat{B}_{i2} \\ \vdots \\ \widehat{B}_{i\alpha_i} \end{bmatrix}, \tag{5.33}
$$

$$
\underset{r_{ik} \times r_{ik}}{J_{ik}} = i \begin{bmatrix} \lambda_i & 1 & & & \\ & \lambda_i & 1 & & \\ & & \ddots & \ddots & \\ & & & \ddots & 1 \\ & & & & \lambda_i \end{bmatrix}, \qquad \underset{r_{ik} \times p}{\widehat{B}_{ik}} = \begin{bmatrix} \widehat{b}_{1ik} \\ \widehat{b}_{2ik} \\ \vdots \\ \widehat{b}_{rik} \end{bmatrix}, \tag{5.34}
$$

If for $i = 1, 2, \ldots, l$, the last row vector of $\widehat{B}_{i1}, \widehat{B}_{i2}, \ldots, \widehat{B}_{i\alpha_i}$ are all pairwise linear independent, which means that:

$$
\mathrm{rank} \begin{bmatrix} \widehat{b}_{ri1} \\ \widehat{b}_{ri2} \\ \vdots \\ \widehat{b}_{ri\alpha_i} \end{bmatrix} = \alpha_i, \quad \forall i = 1, 2, \ldots, l. \tag{5.35}
$$

then the system (5.35) is controllable.

Proof. For simplicity, suppose that:

$$
\widehat{A} = \begin{bmatrix} \lambda_1 & 1 & & & & & \\ & \lambda_1 & 1 & & & & \\ & & \lambda_1 & & & & \\ & & & \lambda_1 & 1 & & \\ & & & & \lambda_1 & & \\ & & & & & \lambda_2 & 1 \\ & & & & & & \lambda_2 \end{bmatrix}, \qquad \widehat{B} = \begin{bmatrix} \widehat{b}_{111} \\ \widehat{b}_{211} \\ \widehat{b}_{r11} \\ \widehat{b}_{112} \\ \widehat{b}_{r12} \\ \widehat{b}_{121} \\ \widehat{b}_{r21} \end{bmatrix}, \tag{5.36}
$$

where $\lambda_1 \neq \lambda_2$. For the above Jordan canonical form, we construct the PBH criteria matrix:

$$
[sI - \widehat{A}, \widehat{B}] = \begin{bmatrix} s - \lambda_1 & -1 & & & & & & \widehat{b}_{111} \\ & s - \lambda_1 & -1 & & & & & \widehat{b}_{211} \\ & & s - \lambda_1 & & & & & \widehat{b}_{r11} \\ & & & s - \lambda_1 & -1 & & & \widehat{b}_{112} \\ & & & & s - \lambda_1 & & & \widehat{b}_{r12} \\ & & & & & s - \lambda_2 & -1 & \widehat{b}_{121} \\ & & & & & & s - \lambda_2 & \widehat{b}_{r21} \end{bmatrix}. \tag{5.37}
$$

We imply the rank criteria when $s = \lambda_1$, which yields:

$$\left[\lambda_1 I - \widehat{A}, \widehat{B}\right] = \begin{bmatrix} 0 & -1 & & & & & & \widehat{b}_{111} \\ & 0 & -1 & & & & & \widehat{b}_{211} \\ & & 0 & & & & & \widehat{b}_{r11} \\ & & & 0 & -1 & & & \widehat{b}_{112} \\ & & & & 0 & & & \widehat{b}_{r12} \\ & & & & & \lambda_1 - \lambda_2 & -1 & \widehat{b}_{121} \\ & & & & & & \lambda_1 - \lambda_2 & \widehat{b}_{r21} \end{bmatrix}, \tag{5.38}$$

where $\lambda_1 - \lambda_2 \neq 0$. Obviously, $[\lambda_1 I - \widehat{A}, \widehat{B}]$ has full rank for the rows; namely, $\mathrm{rank}[\lambda_1 I - \widehat{A}, \widehat{B}] = n = 7$, but only if:

$$\mathrm{rank}\begin{bmatrix} \widehat{b}_{r11} \\ \widehat{b}_{r12} \end{bmatrix} = \alpha_1 = 2. \tag{5.39}$$

Similarly, for $s = \lambda_2$, $[\lambda_1 I - \widehat{A}, \widehat{B}]$ has full rank for rows: $\mathrm{rank}[\lambda_1 I - \widehat{A}, \widehat{B}] = n = 7$, but only if:

$$\mathrm{rank}\,\widehat{b}_{r21} = \alpha_2 = 1. \tag{5.40}$$

Therefore, equation (5.35) is proven. □

5.3.2 Controllable Examples

Example 5.3. Consider the controllability of the following continuous time LTI system:

$$\begin{bmatrix} \dot{x}_1 \\ \dot{x}_2 \end{bmatrix} = \begin{bmatrix} 4 & 0 \\ 0 & -5 \end{bmatrix}\begin{bmatrix} x_1 \\ x_2 \end{bmatrix} + \begin{bmatrix} 1 \\ 2 \end{bmatrix} u, \quad n = 2.$$

Solution. The controllability matrix is:

$$Q_{\mathrm{c}} = \begin{bmatrix} B & AB \end{bmatrix} = \begin{bmatrix} 1 & 4 \\ 2 & -10 \end{bmatrix}.$$

Obviously, rank $Q_{\mathrm{c}} = 2 = n$. According to the rank criteria, the system is controllable.

Example 5.4. Consider the controllability of the following continuous time LTI system:

$$\dot{x} = \begin{bmatrix} -1 & -4 & -2 \\ 0 & 6 & -1 \\ 1 & 7 & -1 \end{bmatrix} x + \begin{bmatrix} 2 & 0 \\ 0 & 1 \\ 1 & 1 \end{bmatrix} u, \quad n = 3.$$

Solution. The controllability matrix is:

$$Q_c = \begin{bmatrix} B & AB & A^2B \end{bmatrix} = \begin{bmatrix} 2 & 0 & -4 & * & * & * \\ 0 & 1 & -1 & * & * & * \\ 1 & 1 & 1 & * & * & * \end{bmatrix}.$$

From the first three columns of Q_c, we can see that:

$$\det \begin{bmatrix} 2 & 0 & -4 \\ 0 & 1 & -1 \\ 1 & 1 & 1 \end{bmatrix} \neq 0$$

$$\text{rank } Q_c = 3 = n.$$

Thus, there is no need to compute the last three columns of Q_c. According to the rank criteria, the system is controllable.

Example 5.5. Consider the controllability of the following continuous time LTI system:

$$\dot{x} = \begin{bmatrix} 0 & 1 & 0 & 0 \\ 0 & 0 & -1 & 0 \\ 0 & 0 & 0 & 1 \\ 0 & 0 & 5 & 0 \end{bmatrix} x + \begin{bmatrix} 0 & 1 \\ 1 & 0 \\ 0 & 1 \\ -2 & 0 \end{bmatrix} u, \quad n = 4.$$

Solution. First, compute the matrix:

$$[sI - A, B] = \begin{bmatrix} s & -1 & 0 & 0 & 0 & 1 \\ 0 & s & 1 & 0 & 1 & 0 \\ 0 & 0 & s & -1 & 0 & 1 \\ 0 & 0 & -5 & s & -2 & 0 \end{bmatrix}.$$

The eigenvalues of A are calculated as:

$$\lambda_1 = \lambda_2 = 0, \quad \lambda_3 = \sqrt{5}, \quad \lambda_4 = -\sqrt{5}.$$

Next, we check the rank of $[sI - A, B]$ for each eigenvalue. For $s = \lambda_1 = \lambda_2 = 0$, we have:

$$\text{rank}[sI - A, B] = \text{rank} \begin{bmatrix} 0 & -1 & 0 & 0 & 0 & 1 \\ 0 & 0 & 1 & 0 & 1 & 0 \\ 0. & 0 & 0 & -1 & 0 & 1 \\ 0 & 0 & -5 & 0 & -2 & 0 \end{bmatrix},$$

$$= \text{rank} \begin{bmatrix} -1 & 0 & 0 & 0 \\ 0 & 1 & 0 & -1 \\ 0 & 0 & -1 & 0 \\ 0 & -5 & 0 & -2 \end{bmatrix} = 4 = n.$$

For $s = \lambda_3 = \sqrt{5}$, we have:

$$\text{rank}[sI - A, B] = \text{rank}\begin{bmatrix} \sqrt{5} & -1 & 0 & 0 & 0 & 1 \\ 0 & \sqrt{5} & 1 & 0 & 1 & 0 \\ 0 & 0 & \sqrt{5} & -1 & 0 & 1 \\ 0 & 0 & -5 & \sqrt{5} & -2 & 0 \end{bmatrix}.$$

$$= \text{rank}\begin{bmatrix} \sqrt{5} & -1 & 0 & 1 \\ 0 & \sqrt{5} & 1 & 0 \\ 0 & 0 & 0 & 1 \\ 0 & 0 & -2 & 0 \end{bmatrix} = 4 = n.$$

For $s = \lambda_4 = -\sqrt{5}$, we have:

$$\text{rank}[sI - A, B] = \text{rank}\begin{bmatrix} -\sqrt{5} & -1 & 0 & 0 & 0 & 1 \\ 0 & -\sqrt{5} & 1 & 0 & 1 & 0 \\ 0 & 0 & -\sqrt{5} & -1 & 0 & 1 \\ 0 & 0 & -5 & -\sqrt{5} & -2 & 0 \end{bmatrix},$$

$$= \text{rank}\begin{bmatrix} -\sqrt{5} & -1 & 0 & 1 \\ 0 & -\sqrt{5} & 1 & 0 \\ 0 & 0 & 0 & 1 \\ 0 & 0 & -2 & 0 \end{bmatrix} = 4 = n.$$

This shows that the given system qualifies the PBH Criteria, and it is controllable.

Example 5.6. Considering a continuous time LTI system with pairwise different eigenvalues. Suppose that the Jordan canonical sate equation is:

$$\begin{bmatrix} \dot{x}_1 \\ \dot{x}_2 \\ \dot{x}_3 \end{bmatrix} = \begin{bmatrix} -7 & 0 & 0 \\ 0 & -2 & 0 \\ 0 & 0 & 1 \end{bmatrix}\begin{bmatrix} x_1 \\ x_2 \\ x_3 \end{bmatrix} + \begin{bmatrix} 0 & 2 \\ 4 & 0 \\ 0 & 1 \end{bmatrix}\begin{bmatrix} u_1 \\ u_2 \end{bmatrix}.$$

Solution. We can see directly that the matrix $\bar{B}$ does not contain zero row vectors. According to Jordan canonical form criteria I, the system is controllable.

Example 5.7. Considering a continuous time LTI system with duplicate eigenvalues. Suppose that the Jordan canonical state equation is:

$$\dot{\hat{x}} = \begin{bmatrix} -2 & 1 & & & & & \\ 0 & -2 & & & & & \\ & & -2 & & & & \\ & & & -2 & & & \\ & & & & 3 & 1 & \\ & & & & 0 & 3 & \\ & & & & & & 3 \end{bmatrix}\hat{x} + \begin{bmatrix} 0 & 0 & 0 \\ 1 & 0 & 0 \\ 0 & 4 & 0 \\ 0 & 0 & 7 \\ 0 & 0 & 0 \\ 1 & 1 & 0 \\ 0 & 4 & 1 \end{bmatrix}u.$$

Considering the last rows of Jordan blocks for $\lambda_1 = -2$ and $\lambda_2 = 3$, we find the corresponding rows in $\hat{B}$ and construct the following two matrices:

$$\begin{bmatrix} \hat{b}_{r11} \\ \hat{b}_{r12} \\ \hat{b}_{r13} \end{bmatrix} = \begin{bmatrix} 1 & 0 & 0 \\ 0 & 4 & 0 \\ 0 & 0 & 7 \end{bmatrix}, \qquad \begin{bmatrix} \hat{b}_{r21} \\ \hat{b}_{r22} \end{bmatrix} = \begin{bmatrix} 1 & 1 & 0 \\ 0 & 4 & 1 \end{bmatrix}.$$

Hence, we can see that both of them have full ranks. According to criteria II of the Jordan canonical form, the system is controllable.

5.3.3 Observable Criteria

Consider the continuous time LTI system:

$$\begin{aligned} \dot{x} &= Ax + Bu \\ y &= Cx + Du \,, \end{aligned} \tag{5.41}$$

where A, B, C, and D are respectively $n \times n$, $n \times p$, $q \times n$, and $q \times p$ constant matrices.

Theorem 5.6 (Gram Matrix Criteria). *The state equation* (5.41) *is observable only if the* $n \times n$ *matrix*

$$W_o[0, t_1] \triangleq \int_0^{t_1} e^{A^T t} C^T C e^{At} dt \tag{5.42}$$

is nonsingular for any $t_1 > 0$.

Proof. We premultiply (5.4) by $e^{A^T t} C^T$, and then integrate it over $[0, t_1]$ to yield:

$$\left(\int_0^{t_1} e^{A^T t} C^T C e^{At} dt \right) x(0) = \int_0^{t_1} e^{A^T t} C^T \overline{y}(t) dt \,. \tag{5.43}$$

If $W_o[0, t_1]$ is nonsingular, then:

$$x(0) = W_o^{-1}[0, t_1] \int_0^{t_1} e^{A^T t} C^T \overline{y}(t) dt \,. \tag{5.44}$$

This yields a unique $x(0)$. It also shows that if $W_o[0, t_1]$ (for any $t_1 > 0$) and is nonsingular, then (5.41) is observable. Next, we show that, if $W_o[0, t_1]$ is singular or, equivalently, positive semidefinite for all t_1, then (5.41) is not observable. If $W_o[0, t_1]$ is positive semidefinite, there exists an $n \times 1$ nonzero constant vector v such that:

$$v^T W_o[0, t_1] v = \int_0^{t_1} v^T e^{A^T t} C^T C e^{At} v dt = \int_0^{t_1} \|C e^{At} v\|^2 dt = 0 \,,$$

which implies that:

$$Ce^{At}v \equiv 0, \tag{5.45}$$

for all t in $[0, t_1]$. If $u \equiv 0$, then $x_1(0) = v \neq 0$ and $x_2(0) = 0$ both yield the same

$$y(t) = Ce^{At}x_i(0) \equiv 0.$$

Two different initial states yield the same zeroinput response. Therefore, we cannot uniquely determine $x(0)$. Thus (5.41) is not observable. This completes the proof for Theorem 5.1. □

Theorem 5.7 (Theorem of Duality). *The pair (A, B) is controllable only if the pair (A^T, B^T) is observable.*

Proof. The pair (A, B) is controllable only if

$$W_c[0, t_1] = \int_0^{t_1} e^{At}BB^T e^{A^T t} dt$$

is nonsingular for any t_1. The pair (A^T, B^T) is observable only if, by replacing A with A^T and C with B^T in (5.42), $W_0[0, t_1] = \int_0^{t_1} e^{At}BB^T e^{A^T t} dt$ is nonsingular for any t. □

Theorem 5.8 (Rank Criteria). *The state equation (5.41) is observable only if the $nq \times n$ observability matrix*

$$Q_o = \begin{bmatrix} C \\ CA \\ \vdots \\ CA^{n-1} \end{bmatrix}, \quad or \quad Q_o^T = \begin{bmatrix} C^T & A^T C^T & \cdots & (A^T)^{n-1} C^T \end{bmatrix},$$

has rank n.

Theorem 5.9 (Observable PBH Rank Criteria). *The state equation (5.41) is observable only if:*

$$\text{rank} \begin{bmatrix} \lambda_i I - A \\ C \end{bmatrix} = n, \quad and \quad i = 1, 2, \ldots, n,$$

at the eigenvalue, λ, of A.

Theorem 5.10 (Observable PBH Characteristic Vector Criteria). *The state equation (5.41) is observable only if an orthogonal nonzero right characteristic vector for all rows of matrix C in matrix A does not exist. Equivalently, the only right characteristic vector at every eigenvalue λ, of A, that can satisfy the following equations:*

$$A\bar{\alpha} = \lambda_i \bar{\alpha}, \quad C\bar{\alpha} = 0,$$

is $\bar{\alpha} = 0$.

Theorem 5.11 (Observable Jordan Canonical Form Criteria). *Assuming $\lambda_i \neq \lambda_j$, $\forall i \neq j$, the eigenvalues of system (5.41) are λ_i (σ_i layers, α_i layers) with i varying from 1 to l, and $(\sigma_1 + \sigma_2 + \ldots \sigma_l) = n$. The Jordan canonical form of the system is obtained by linearly nonsingular transformation:*

$$\dot{\hat{x}} = \widehat{A}\hat{x}$$
$$y = \widehat{C}\hat{x},$$

where:

$$\underset{(n\times n)}{\widehat{A}} = \begin{bmatrix} J_1 & & & \\ & J_2 & & \\ & & \ddots & \\ & & & J_l \end{bmatrix}, \qquad \underset{(q\times n)}{\widehat{C}} = \begin{bmatrix} \widehat{C}_1 & \widehat{C}_2 & \cdots & \widehat{C}_l \end{bmatrix},$$

$$\underset{(\sigma_i\times\sigma_i)}{J_i} = \begin{bmatrix} J_{i1} & & & \\ & J_{i2} & & \\ & & \ddots & \\ & & & J_{i\alpha_i} \end{bmatrix}, \qquad \underset{(q\times\sigma_i)}{\widehat{C}_i} = \begin{bmatrix} \widehat{C}_{i1} & \widehat{C}_{i2} & \cdots & \widehat{C}_{i\alpha_i} \end{bmatrix},$$

$$\underset{(r_{ik}\times r_{ik})}{J_{ik}} = \begin{bmatrix} \lambda_i & 1 & & & \\ & \lambda_i & 1 & & \\ & & \ddots & \ddots & \\ & & & \ddots & 1 \\ & & & & \lambda_i \end{bmatrix}, \qquad \underset{(q\times r_{ik})}{\widehat{C}_{ik}} = \begin{bmatrix} \widehat{c}_{1ik} & \widehat{c}_{2ik} & \cdots & \widehat{c}_{rik} \end{bmatrix}.$$

For $i = 1, 2, \ldots, l$, the first columns of $\widehat{C}_{i1}, \widehat{C}_{i2}, \ldots, \widehat{C}_{i\alpha_i}$ are linearly independent. That is:

$$\text{rank}\begin{bmatrix} \widehat{c}_{1i1} & \widehat{c}_{1i2} & \cdots & \widehat{c}_{1i\alpha_i} \end{bmatrix} = \alpha_i, \quad \forall i = 1, 2, \ldots, l.$$

5.3.4 Observable Examples

Example 5.8. Is the state equation:

$$\dot{x} = \begin{bmatrix} -1 & -4 & -2 \\ 0 & 6 & -1 \\ 1 & 7 & -1 \end{bmatrix} x, \quad n = 3,$$

$$y = \begin{bmatrix} 0 & 2 & 1 \\ 1 & 1 & 0 \end{bmatrix} x$$

observable?

Solution.

$$\text{rank } Q_o = \text{rank} \begin{bmatrix} C \\ CA \\ CA^2 \end{bmatrix} = \text{rank} \begin{bmatrix} 0 & 2 & 1 \\ 1 & 1 & 0 \\ 1 & 19 & -3 \\ * & * & * \\ * & * & * \\ * & * & * \end{bmatrix} = 3 = n \, .$$

Obviously, we know that the matrix Q_o has full rank from the first three rows. Therefore, the system is completely observable from *rank criteria*.

Example 5.9. Is the state equation:

$$\dot{x} = \begin{bmatrix} 0 & 1 & 0 & 0 \\ 0 & 0 & -1 & 0 \\ 0 & 0 & 0 & 1 \\ 0 & 0 & 5 & 0 \end{bmatrix} x \, , \quad n = 4 \, ,$$

$$y = \begin{bmatrix} 0 & 1 & 0 & -2 \\ 1 & 0 & 1 & 0 \end{bmatrix} x$$

observable?

Solution. First we compute the eigenvalues of A:

$$|\lambda I - A| = \begin{vmatrix} \lambda & -1 & 0 & 0 \\ 0 & \lambda & 1 & 0 \\ 0 & 0 & \lambda & -1 \\ 0 & 0 & -5 & \lambda \end{vmatrix} = 0 \, ,$$

$$\lambda_1 = \lambda_2 = 0 \, , \quad \lambda_3 = \sqrt{5} \, , \quad \lambda_3 = -\sqrt{5} \, .$$

According to *observable PBH rank criteria*, we compute:

$$\text{rank} \begin{bmatrix} \lambda I - A \\ C \end{bmatrix}_{\lambda=0} = \text{rank} \begin{bmatrix} 0 & -1 & 0 & 0 \\ 0 & 0 & 1 & 0 \\ 0 & 0 & 0 & -1 \\ 0 & 0 & -5 & 0 \\ 0 & 1 & 0 & -2 \\ 1 & 0 & 1 & 0 \end{bmatrix} = 4 = n \, ,$$

$$\text{rank} \begin{bmatrix} \lambda I - A \\ C \end{bmatrix}_{\lambda=\sqrt{5}} = \text{rank} \begin{bmatrix} \sqrt{5} & -1 & 0 & 0 \\ 0 & \sqrt{5} & 1 & 0 \\ 0 & 0 & \sqrt{5} & -1 \\ 0 & 0 & -5 & \sqrt{5} \\ 0 & 1 & 0 & -2 \\ 1 & 0 & 1 & 0 \end{bmatrix} = 4 = n \, ,$$

$$\text{rank}\begin{bmatrix}\lambda I - A \\ C\end{bmatrix}_{\lambda=-\sqrt{5}} = \text{rank}\begin{bmatrix} -\sqrt{5} & -1 & 0 & 0 \\ 0 & -\sqrt{5} & 1 & 0 \\ 0 & 0 & -\sqrt{5} & -1 \\ 0 & 0 & -5 & -\sqrt{5} \\ 0 & 1 & 0 & -2 \\ 1 & 0 & 1 & 0 \end{bmatrix} = 4 = n,$$

which satisfy *the observable PBH rank criteria*. Therefore, the system is completely observable.

Example 5.10. Consider a system with pairwise different eigenvalues, and suppose its Jordan canonical form is:

$$\begin{bmatrix} \dot{x}_1 \\ \dot{x}_2 \\ \dot{x}_3 \end{bmatrix} = \begin{bmatrix} -7 & 0 & 0 \\ 0 & -2 & 0 \\ 0 & 0 & 1 \end{bmatrix}\begin{bmatrix} x_1 \\ x_2 \\ x_3 \end{bmatrix},$$

$$y = \begin{bmatrix} 0 & 4 & 0 \\ 2 & 0 & 1 \end{bmatrix}x.$$

Examine the observability of the system.

Solution. We know the matrix $\bar{C}$ does not consist of a column whose elements are all zero. According to the *observable Jordan canonical form criteria*, we know that the system is observable.

Example 5.11. Consider an LTI system with duplicate eigenvalues. Suppose that the Jordan canonical form is as follows:

$$\dot{x} = \begin{bmatrix} -2 & 1 & & & & & \\ 0 & -2 & & & & & \\ & & -2 & & & & \\ & & & -2 & & & \\ & & & & 3 & 1 & \\ & & & & 0 & 3 & \\ & & & & & & 3 \end{bmatrix}\hat{x},$$

$$y = \begin{bmatrix} 1 & 0 & 0 & 0 & 1 & 0 & 0 \\ 0 & 0 & 4 & 0 & 1 & 0 & 4 \\ 0 & 0 & 0 & 7 & 0 & 0 & 1 \end{bmatrix}\hat{x}.$$

Solution. Consider the first column of two Jordan blocks for $\lambda = -2$ and $\lambda = 3$. Find the corresponding columns from the matrix $\hat{C}$ and construct the following two matrices:

$$\begin{bmatrix} \hat{c}_{111} & \hat{c}_{112} & \hat{c}_{113} \end{bmatrix} = \begin{bmatrix} 1 & 0 & 0 \\ 0 & 4 & 0 \\ 0 & 0 & 7 \end{bmatrix},$$

$$\begin{bmatrix} \hat{c}_{121} & \hat{c}_{122} \end{bmatrix} = \begin{bmatrix} 1 & 0 \\ 1 & 4 \\ 0 & 1 \end{bmatrix}.$$

We know that the two matrices are both linearly independent columns. According to the *observable Jordan canonical form criteria*, we know that the system is observable.

5.4 Duality System

5.4.1 Definition

There are two systems. The system Σ_1 is shown as follows:

$$\dot{x}_1 = A_1 x_1 + B_1 u_1$$
$$y_1 = C_1 x_1 .$$

The other system, Σ_2, is:

$$\dot{x}_2 = A_2 x_2 + B_2 u_2$$
$$y_2 = C_2 x_2 .$$

If the following conditions are satisfied, the system Σ_1 and Σ_2 are duality systems.

$$A_2 = A_1^{\mathrm{T}}, \quad B_2 = C_1^{\mathrm{T}}, \quad C_2 = B_1^{\mathrm{T}}, \tag{5.46}$$

where x_1, x_2 are n-dimensional state vectors, u_1, u_2 are r-dimensional and m-dimensional control vectors respectively; y_1, y_2 are m-dimensional and r-dimensional output vectors respectively; A_1, A_2 are $n \times n$ system matrices; A_1, A_2 are $n \times r$ and $n \times m$ control matrices respectively, and C_1, C_2 are $m \times n$ and $r \times n$ output matrices respectively.

Obviously, the system Σ_1 is an n-order system with r inputs and m outputs, while its duality systems Σ_2 is an n-order system with m inputs and r outputs. The configurations of the duality systems Σ_1 and Σ_2 are shown in Figure 5.4.

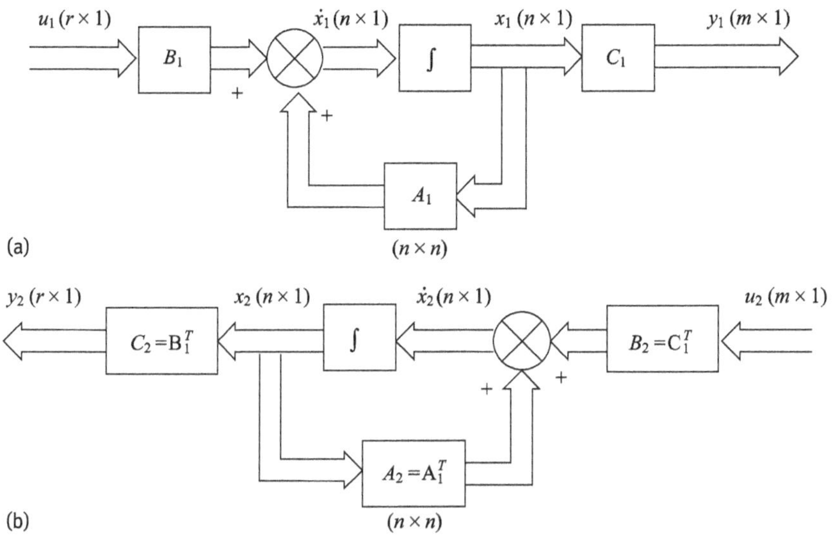

Fig. 5.4: Configuration diagrams of duality systems.

5.4.2 Properties of Duality Systems

Conclusion 5.1. *No matter whether a system is a continuous time system or a discrete time system, if the system Σ_1 is linear, its duality system Σ_2 is also linear. Similarly, if the system Σ_1 is time variable or time invariable, its duality system Σ_2 is also time variable or time invariable.*

Conclusion 5.2. *The transfer function matrices of the duality systems are mutually inverted.*

Proof. As shown in Figure 5.4 (a), the transfer function matrix $W_1(s)$ of the system Σ_1 is a $m \times r$ matrix as follows:

$$W_1(s) = C_1(sI - A_1)^{-1}B_1 \ .$$

As shown in Figure 5.4 (b), the transfer function matrix $W_2(s)$ of the system Σ_2 is a $r \times m$ matrix as follows:

$$
\begin{aligned}
W_2(s) &= C_2(sI - A_2)^{-1}B_2 \\
&= B_1^T(sI - A_1^T)^{-1}C_1^T \\
&= B_1^T\left[(sI - A_1)^{-1}\right]^T C_1^T \ .
\end{aligned}
$$

Obviously,

$$[W_2(s)]^T = C_1(sI - A_1)^{-1}B_1 = W_1(s) \ .$$

In the same way we can know that the input state transfer function matrix $(sI - A_1)^{-1}B_1$ of the system Σ_1 and state input transfer function matrix $C_2(sI - A_2)^{-1}$ of the system Σ_2 are mutually inverted. The state input transfer function matrix $C_1(sI - A_1)^{-1}$ of the system Σ_1 and input state transfer function matrix $(sI - A_2)^{-1}B_2$ of the system Σ_2 are mutually inverted.

In addition, the characteristic equations of duality systems are the same. That is to say:

$$|sI - A_2| = |sI - A_1^{\mathrm{T}}| = |sI - A_1| .$$ □

Conclusion 5.3. *The system $\Sigma_1 = (A_1, B_1, C_1)$ and the system $\Sigma_2 = (A_2, B_2, C_2)$ are duality systems. The controllability of system Σ_1 is equivalent to the observability of system Σ_2. Similarly, the observability of system Σ_1 is equivalent to the controllability of system Σ_2. In other words, if the system Σ_1 is controllable or observable, the system Σ_2 is observable or controllable.*

Proof. The controllability matrix of system Σ_2 is:

$$Q_{2c} = \begin{bmatrix} B_2 & A_2 B_2 & \cdots & A_2^{n-1} B_2 \end{bmatrix} ,$$

and its rank equals n. Therefore, the system Σ_2 is controllable.

The integration of the equation (5.46) in the above equation results in the following equation:

$$Q_{2c} = \begin{bmatrix} C_1^{\mathrm{T}} & A_1^{\mathrm{T}} C_1^{\mathrm{T}} & \cdots & (A_1^{\mathrm{T}})^{(n-1)} C_1^{\mathrm{T}} \end{bmatrix} = Q_{1o}^{\mathrm{T}} ,$$

where Q_{1o} is the observability matrix of system Σ_1.

This indicates the rank of the observability matrix of system Σ_1 is n. Therefore, system Σ_1 is observable.

In the same way, we know:

$$\begin{aligned} Q_{2o}^{\mathrm{T}} &= \begin{bmatrix} C_2^{\mathrm{T}} & A_2^{\mathrm{T}} C_2^{\mathrm{T}} & \cdots & (A_2^{\mathrm{T}})^{n-1} C_2^{\mathrm{T}} \end{bmatrix} \\ &= \begin{bmatrix} B_1 & A_1 B_1 & \cdots & A_1^{n-1} B_1 \end{bmatrix} = Q_{1c} . \end{aligned}$$

If Q_{2o} has full rank, the system Σ_2 is observable, and Q_{1c} also has full rank. Therefore, the system Σ_1 is controllable. □

5.5 Canonical Form

This section discusses the controllability canonical forms and observability canonical forms of state equations.

5.5.1 Controllability Canonical Form of Single Input Systems

Consider the n-dimensional time invariant system:

$$\dot{x} = Ax + Bu$$
$$y = Cx .$$

If the system is controllable, we have:

$$\text{rank}\begin{bmatrix} B & AB & \ldots & A^{n-1}B \end{bmatrix} = n .$$

Hence, there are at least n linearly independent n-dimensional column vectors in the controllability matrix. We select n linearly independent vectors from the nr column vectors and make some linear transformations. Then we can get a certain controllability canonical form, whose columns are still linear independent. For single input single output systems, there is only one group of linearly independent vectors, so the controllability canonical form is unique. However, for multiple input multiple output systems, there are various choices of n linearly independent vectors, so the controllability canonical form is not unique. Obviously, the canonical forms exist if, and only if, the system is controllable.

Controllability Canonical Form I

If the LTI single input system:

$$\dot{x} = Ax + bu$$
$$y = Cx \tag{5.47}$$

is controllable, then there exists a linear nonsingular transformation:

$$x = T_{c1}\overline{x} ,$$

$$T_{c1} = \begin{bmatrix} A^{n-1}b & A^{n-2}b & \ldots & b \end{bmatrix} \begin{bmatrix} 1 & & & & 0 \\ \alpha_{n-1} & 1 & & & \\ \vdots & & \ddots & & \\ \alpha_2 & \alpha_3 & & \ddots & \\ \alpha_1 & \alpha_2 & \ldots & \alpha_{n-1} & 1 \end{bmatrix} , \tag{5.48}$$

which can transfer the state space equation into:

$$\dot{\overline{x}} = \overline{A}\overline{x} + \overline{b}u$$
$$y = \overline{C}\overline{x} , \tag{5.49}$$

where:

$$\overline{A} = T_{c1}^{-1}AT_{c1} = \begin{bmatrix} 0 & 1 & & \\ \vdots & & \ddots & \\ 0 & & & 1 \\ -\alpha_0 & -\alpha_1 & \ldots & -\alpha_{n-1} \end{bmatrix} , \quad \overline{b} = T_{c1}^{-1}b = \begin{bmatrix} 0 \\ 0 \\ \vdots \\ 1 \end{bmatrix} , \tag{5.50}$$

$$\overline{C} = CT_{c1} = \begin{bmatrix} \beta_0 & \beta_1 & \ldots & \beta_{n-1} \end{bmatrix} . \tag{5.51}$$

Equation (5.49) is called the controllability canonical form I, where α_i ($i = 0, 1, \ldots, n-1$) are the coefficients of the following polynomial:

$$|\lambda I - A| = \lambda^n + \alpha_{n-1}\lambda^{n-1} + \cdots + \alpha_1\lambda + \alpha_0 .$$

β_i ($i = 0, 1, \ldots, n-1$) are the results of CT_{c1}:

$$
\left.
\begin{aligned}
\beta_0 &= C(A^{n-1}b + \alpha_{n-1}A^{n-2}b + \cdots + \alpha_1 b) \\
&\vdots \\
\beta_{n-2} &= C(Ab + \alpha_{n-1}b) \\
\beta_{n-1} &= Cb
\end{aligned}
\right\}
\tag{5.52}
$$

Proof. Suppose the system is controllable; the $n \times 1$ vectors b, Ab, $\ldots$, $A^{n-1}b$ are linearly independent, and the new vectors e_1, e_2, $\ldots$, e_n in the following combination are also linearly independent:

$$
\left.
\begin{aligned}
e_1 &= A^{n-1}b + \alpha_{n-1}A^{n-2}b + \alpha_{n-2}A^{n-3}b + \cdots + \alpha_1 b \\
e_2 &= A^{n-2}b + \alpha_{n-1}A^{n-3}b + \cdots + \alpha_2 b \\
&\vdots \\
e_{n-1} &= Ab + \alpha_{n-1}b \\
e_n &= b,
\end{aligned}
\right\}
\tag{5.53}
$$

where α_i ($i = 0, 1, \ldots, n-1$) are the coefficients of the polynomial.

Thus, the transformation matrix T_{c1} are composed of $e_1, e_2, \ldots, e_n$:

$$
T_{c1} = \begin{bmatrix} e_1 & e_2 & \cdots & e_n \end{bmatrix}.
\tag{5.54}
$$

As $\overline{A} = T_{c1}^{-1} A T_{c1}$, we have:

$$
T_{c1}\overline{A} = AT_{c1} = A\begin{bmatrix} e_1 & e_2 & \cdots & e_n \end{bmatrix} = \begin{bmatrix} Ae_1 & Ae_2 & \cdots & Ae_n \end{bmatrix}.
\tag{5.55}
$$

Substituting equation (5.53) into the above equation yields:

$$
\begin{aligned}
Ae_1 &= A(A^{n-1}b + \alpha_{n-1}A^{n-2}b + \cdots + \alpha_1 b) \\
&= (A^n b + \alpha_{n-1}A^{n-1}b + \cdots + \alpha_1 Ab + \alpha_0 b) - \alpha_0 b \\
&= -\alpha_0 b = -\alpha_0 e_n \\
Ae_2 &= A(A^{n-2}b + \alpha_{n-1}A^{n-3}b + \cdots + \alpha_2 b) \\
&= (A^{n-1}b + \alpha_{n-1}A^{n-2}b + \cdots + \alpha_2 Ab + \alpha_1 b) - \alpha_1 b \\
&= e_1 - \alpha_1 e_n \\
&\vdots \\
Ae_{n-1} &= A(Ab + \alpha_{n-1}b) \\
&= (A^2 b + \alpha_{n-1}Ab + \alpha_{n-2}b) - \alpha_{n-2}b \\
&= e_{n-2} - \alpha_{n-2}e_n \\
Ae_n &= Ab = (Ab + \alpha_{n-1}b) - \alpha_{n-1}b = e_{n-1} - \alpha_{n-1}e_n.
\end{aligned}
$$

Then we substitute $Ae_1, Ae_2, \ldots, Ae_n$ into (5.55):

$$T_{c1}\overline{A} = \begin{bmatrix} Ae_1 & Ae_2 & \cdots & Ae_n \end{bmatrix} = \begin{bmatrix} -\alpha_0 e_n & (e_1 - \alpha_1 e_n) & \cdots & (e_{n-1} - \alpha_{n-1}e_n) \end{bmatrix}$$

$$= \begin{bmatrix} e_1 & e_2 & \cdots & e_n \end{bmatrix} \begin{bmatrix} 0 & 1 & & \\ \vdots & & \ddots & \\ 0 & & & 1 \\ -\alpha_0 & -\alpha_1 & \cdots & -\alpha_{n-1} \end{bmatrix}.$$

Furthermore, we deduce $\overline{b}$. From $\overline{b} = T_{c1}^{-1}b$, we have $T_{c1}\overline{b} = b$. Substituting $b = e_n$ yields:

$$T_{c1}\overline{b} = e_n = \begin{bmatrix} e_1 & e_2 & \cdots & e_n \end{bmatrix} \begin{bmatrix} 0 \\ 0 \\ \vdots \\ 1 \end{bmatrix}.$$

Therefore,

$$\overline{b} = \begin{bmatrix} 0 \\ 0 \\ \vdots \\ 1 \end{bmatrix}.$$

Finally, we deduce $\overline{C}$. From $\overline{C} = CT_{c1}$, we get:

$$\overline{C} = CT_{c1} = C \begin{bmatrix} e_1 & e_2 & \cdots & e_n \end{bmatrix}.$$

Substituting (5.53) into the above equation yields:

$$\overline{C} = C \begin{bmatrix} A^{n-1}b + \alpha_{n-1}A^{n-2}b + \alpha_{n-2}A^{n-3}b + \cdots + \alpha_1 b & \cdots & Ab + \alpha_{n-1}b & b \end{bmatrix}$$

$$= \begin{bmatrix} \beta_0 & \beta_1 & \cdots & \beta_{n-1} \end{bmatrix},$$

where:

$$\beta_0 = C(A^{n-1}b + \alpha_{n-1}A^{n-2}b + \alpha_{n-2}A^{n-3}b + \cdots + \alpha_1 b)$$

$$\vdots$$

$$\beta_{n-2} = C(Ab + \alpha_{n-1}b)$$

$$\beta_{n-1} = Cb.$$

We can derive the transfer function easily from the controllability canonical form I:

$$W(s) = \overline{C}(sI - \overline{A})^{-1}\overline{b} = \frac{\beta_{n-1}s^{n-1} + \beta_{n-2}s^{n-2} + \cdots + \beta_1 s + \beta_0}{s^n + \alpha_{n-1}s^{n-1} + \cdots + \alpha_1 s + \alpha_0}. \qquad (5.56)$$

From (5.56), we can see that the coefficients of the denominator polynomial are the negative value of the elements of the last row of $\overline{A}$, and that the coefficients of the numerator polynomial are the elements of $\overline{C}$. Hence, we can write out $\overline{A}, \overline{b}, \overline{C}$, directly from the coefficients of the denominator polynomial and numerator polynomial of the system transfer function. □

Controllability Canonical Form II

If the LTI single input system,

$$\dot{x} = Ax + bu$$
$$y = Cx,$$

(5.57)

is controllable, then there exists a linear nonsingular transformation:

$$x = T_{c2}\overline{x} = \begin{bmatrix} b & Ab & \cdots & A^{n-1}b \end{bmatrix} \overline{x},$$

(5.58)

which will transfer the state space equation into:

$$\dot{\overline{x}} = \overline{A}\overline{x} + \overline{b}u$$
$$y = \overline{C}\overline{x}.$$

(5.59)

Here:

$$\overline{A} = T_{c2}^{-1}AT_{c2} = \begin{bmatrix} 0 & 0 & \cdots & 0 & -\alpha_0 \\ 1 & 0 & \cdots & 0 & -\alpha_1 \\ 0 & 1 & \cdots & 0 & -\alpha_2 \\ \vdots & \vdots & & 0 & \vdots \\ 0 & 0 & \cdots & 1 & -\alpha_{n-1} \end{bmatrix}, \quad \overline{b} = T_{c2}^{-1}b = \begin{bmatrix} 1 \\ 0 \\ \vdots \\ 0 \end{bmatrix}.$$

(5.60)

$$\overline{C} = CT_{c2} = \begin{bmatrix} \beta_0 & \beta_1 & \cdots & \beta_{n-1} \end{bmatrix}$$

(5.61)

Equation (5.59) is called the controllability canonical form II, where $\alpha_0, \alpha_1, \ldots, \alpha_{n-1}$ are the coefficients of the following polynomial:

$$|\lambda I - A| = \lambda^n + \alpha_{n-1}\lambda^{n-1} + \cdots + \alpha_1\lambda + \alpha_0.$$

$\beta_0, \beta_1, \ldots, \beta_{n-1}$ are the results of CT_{c2}:

$$\left. \begin{array}{l} \beta_0 = Cb \\ \beta_1 = CAb \\ \vdots \\ \beta_{n-1} = CA^{n-1}b \end{array} \right\}$$

(5.62)

Proof. As the system is controllable, the controllability matrix

$$Q_c = \begin{bmatrix} b & Ab & \cdots & A^{n-1}b \end{bmatrix}$$

is nonsingular. Let:

$$x = T_{c2}\overline{x}$$

where $T_{c2} = \begin{bmatrix} b & Ab & \cdots & A^{n-1}b \end{bmatrix}$. Thus the state equation and output equation after transformation is:

$$\dot{\overline{x}} = \overline{A}\overline{x} + \overline{b}u = T_{c2}^{-1}AT_{c2}\overline{x} + T_{c2}^{-1}bu$$
$$y = \overline{C}\overline{x} = CT_{c2}\overline{x}.$$

First, we deduce $\overline{A}$:

$$AT_{c2} = A \begin{bmatrix} b & Ab & \cdots & A^{n-1}b \end{bmatrix} = \begin{bmatrix} Ab & A^2b & \cdots & A^nb \end{bmatrix} . \tag{5.63}$$

According to the Cayley–Hamilton theorem, we have:

$$A^n = -\alpha_{n-1}A^{n-1} - \alpha_{n-2}A^{n-2} - \cdots - \alpha_1 A - \alpha_0 .$$

Substituting the above equation into (5.63) yields:

$$AT_{c2} = \begin{bmatrix} Ab & A^2b & \cdots & (-\alpha_{n-1}A^{n-1} - \alpha_{n-2}A^{n-2} - \cdots - \alpha_0)b \end{bmatrix} . \tag{5.64}$$

Now, rewrite (5.64) into matrix form:

$$AT_{c2} = \begin{bmatrix} b & Ab & \cdots & A^{n-1}b \end{bmatrix} \begin{bmatrix} 0 & 0 & \cdots & 0 & -\alpha_0 \\ 1 & 0 & \cdots & 0 & -\alpha_1 \\ 0 & 1 & \cdots & 0 & -\alpha_2 \\ \vdots & \vdots & & 0 & \vdots \\ 0 & 0 & \cdots & 1 & -\alpha_{n-1} \end{bmatrix} ,$$

thus arriving at:

$$AT_{c2} = T_{c2} \begin{bmatrix} 0 & 0 & \cdots & 0 & -\alpha_0 \\ 1 & 0 & \cdots & 0 & -\alpha_1 \\ 0 & 1 & \cdots & 0 & -\alpha_2 \\ \vdots & \vdots & & 0 & \vdots \\ 0 & 0 & \cdots & 1 & -\alpha_{n-1} \end{bmatrix} .$$

Then multiply the above equation by T_{c2}^{-1}:

$$\overline{A} = T_{c2}^{-1} AT_{c2} = \begin{bmatrix} 0 & 0 & \cdots & 0 & -\alpha_0 \\ 1 & 0 & \cdots & 0 & -\alpha_1 \\ 0 & 1 & \cdots & 0 & -\alpha_2 \\ \vdots & \vdots & & 0 & \vdots \\ 0 & 0 & \cdots & 1 & -\alpha_{n-1} \end{bmatrix} . \tag{5.65}$$

As $\overline{b} = T_{c2}^{-1}b$, equally $b = T_{c2}\overline{b} = \begin{bmatrix} b & Ab & \cdots & A^{n-1}b \end{bmatrix}$. Obviously, in order to guarantee (5.65), $\overline{b}$ should satisfy:

$$\overline{b} = \begin{bmatrix} 1 \\ 0 \\ \vdots \\ 0 \end{bmatrix} ,$$

$$\overline{C} = CT_{c2} = \begin{bmatrix} Cb & CAb & \cdots & CA^{n-1}b \end{bmatrix} .$$

That is:

$$\overline{C} = CT_{c2} = \begin{bmatrix} \beta_0 & \beta_1 & \cdots & \beta_{n-1} \end{bmatrix} . \qquad\qquad \square$$

5.5.2 Observability Canonical Form of Single Output Systems

Similar to the condition for controllability canonical forms, the system has observability canonical form only when it is observable. That is:

$$\text{rank} \begin{bmatrix} C \\ CA \\ \vdots \\ CA^{n-1} \end{bmatrix} = n .$$

The state space equation has two types of observability canonical forms: observability canonical form I and observability canonical form II.

Observability Canonical Form I

If the LTI single output system

$$\dot{x} = Ax + bu$$
$$y = Cx \tag{5.66}$$

is controllable, then there exists a linear nonsingular transformation:

$$x = T_{o1}\tilde{x} , \tag{5.67}$$

which will transfer the state space equation (5.65) into:

$$\dot{\tilde{x}} = \tilde{A}\tilde{x} + \tilde{b}u$$
$$y = \tilde{C}\tilde{x} . \tag{5.68}$$

Define the inverse of the transformation matrix T_{o1}:

$$T_{o1}^{-1} = N = \begin{bmatrix} C \\ CA \\ \vdots \\ CA^{n-1} \end{bmatrix} . \tag{5.69}$$

Here:

$$\tilde{A} = T_{o1}^{-1}AT_{o1} = \begin{bmatrix} 0 & 1 & 0 & \cdots & 0 \\ 0 & 0 & 1 & \cdots & 0 \\ 0 & 0 & 0 & \cdots & 0 \\ \vdots & \vdots & \vdots & & \vdots \\ -\alpha_0 & -\alpha_1 & -\alpha_2 & \cdots & -\alpha_{n-1} \end{bmatrix}, \quad \tilde{b} = T_{o1}^{-1}b = \begin{bmatrix} \beta_0 \\ \beta_1 \\ \vdots \\ \beta_{n-1} \end{bmatrix}, \tag{5.70}$$

$$\tilde{C} = CT_{o1} = \begin{bmatrix} 1 & 0 & \cdots & 0 \end{bmatrix} . \tag{5.71}$$

Equation (5.71) is called the observability canonical form I, where α_i ($i = 0, 1, \ldots, n - 1$) are the coefficients of the polynomial A.

Observability Canonical Form II

If the LTI single output system:

$$\dot{x} = Ax + bu$$
$$y = Cx$$

(5.72)

is controllable, then there exists a linear nonsingular transformation:

$$x = T_{02}\tilde{x},$$

(5.73)

where:

$$T_{02} = \begin{bmatrix} 1 & \alpha_{n-1} & \cdots & \alpha_2 & \alpha_1 \\ 0 & 1 & \cdots & \alpha_3 & \alpha_2 \\ \vdots & \vdots & & \vdots & \vdots \\ 0 & 0 & \cdots & 1 & \alpha_{n-1} \\ 0 & 0 & \cdots & 0 & 1 \end{bmatrix} \begin{bmatrix} CA^{n-1} \\ CA^{n-2} \\ \vdots \\ CA \\ C \end{bmatrix}.$$

(5.74)

The state space equation after transformation is:

$$\dot{\tilde{x}} = \tilde{A}\tilde{x} + \tilde{b}u$$
$$y = \tilde{C}\tilde{x},$$

(5.75)

where:

$$\tilde{A} = T_{02}^{-1}AT_{02} = \begin{bmatrix} 0 & 0 & \cdots & 0 & -\alpha_0 \\ 1 & 0 & \cdots & 0 & -\alpha_1 \\ 0 & 1 & 0 & \vdots & -\alpha_2 \\ \vdots & \vdots & & 0 & \vdots \\ 0 & 0 & \cdots & 1 & -\alpha_{n-1} \end{bmatrix}, \quad \tilde{b} = T_{02}^{-1}b = \begin{bmatrix} \beta_0 \\ \beta_1 \\ \vdots \\ \beta_{n-1} \end{bmatrix},$$

(5.76)

$$\tilde{C} = CT_{01} = \begin{bmatrix} 0 & 0 & \cdots & 1 \end{bmatrix}.$$

(5.77)

Equation (5.75) is called the observability canonical form II, where α_i ($i = 0, 1, \ldots,$ $n-1$) are the coefficients of the polynomial A, and β_i ($i = 0, 1, \ldots, n-1$) are the results of $T_{02}^{-1}b$ and is shown in (5.51).

The observability canonical form I is dual to the controllability canonical form II. And the observability canonical form II is dual to the controllability canonical form I.

Proof. We prove it by the theory of duality.

First we construct the duality system of $\Sigma = (A, b, C)$:

$$A^* = A^T$$
$$b^* = C^T$$
$$C^* = b^T.$$

This gives us the controllability canonical form II of $\Sigma^* = (A^*, b^*, C^*)$. The observability canonical form I of Σ is just the controllability canonical form II of Σ^*. For example:

$$\widetilde{A} = A^* = \overline{A}^{\mathrm{T}}$$

$$\widetilde{b} = b^* = \overline{C}^{\mathrm{T}}$$

$$\widetilde{C} = C^* = \overline{b}^{\mathrm{T}},$$

where:

$\overline{A}, \overline{b}, \overline{C}$ the coefficient matrices of the controllability canonical form II of Σ;

$\widetilde{A}, \widetilde{b}, \widetilde{C}$ the coefficient matrices of the observability canonical form I of Σ;

A^*, b^*, C^* the coefficient matrices of the controllability canonical form II of Σ^*.

□

Therefore the transformation above can be deduced directly by the theory of duality. The same is true for observability canonical form I. Besides, we can also derive the transfer function directly from the controllability canonical form II:

$$W(s) = \frac{\beta_{n-1}s^{n-1} + \beta_{n-2}s^{n-2} + \cdots + \beta_0}{s^n + \alpha_{n-1}s^{n-1} + \alpha_{n-2}s^{n-2} \cdots + \alpha_0}, \tag{5.78}$$

Where the coefficients of the denominator polynomial are the negative value of the elements of the last column of $\widetilde{A}$, and the coefficients of the numerator polynomial are the elements of $\widetilde{b}$.

5.5.3 Examples

Example 5.12. Try to transform the following state space equation into controllability canonical form I:

$$\dot{x} = \begin{bmatrix} 1 & 2 & 0 \\ 3 & -1 & 1 \\ 0 & 2 & 0 \end{bmatrix} x + \begin{bmatrix} 2 \\ 1 \\ 1 \end{bmatrix} u,$$

$$y = \begin{bmatrix} 0 & 0 & 1 \end{bmatrix} x.$$

Solution. First we judge the controllability of the system:

$$Q_c = \begin{bmatrix} b & Ab & A^2b \end{bmatrix} = \begin{bmatrix} 2 & 4 & 16 \\ 1 & 6 & 8 \\ 1 & 2 & 12 \end{bmatrix}.$$

rank $Q_c = 3$ so the system is controllable.

Then we compute the characteristic polynomial:

$$|\lambda I - A| = \lambda^3 - 9\lambda + 2.$$

Thus, $\alpha_2 = 0$, $\alpha_1 = -9$, $\alpha_0 = 2$. From equations (5.49) and (5.50), we have:

$$\bar{A} = \begin{bmatrix} 0 & 1 & 0 \\ 0 & 0 & 1 \\ -\alpha_0 & -\alpha_1 & -\alpha_2 \end{bmatrix} = \begin{bmatrix} 0 & 1 & 0 \\ 0 & 0 & 1 \\ -2 & 9 & 0 \end{bmatrix},$$

$$\bar{C} = C\begin{bmatrix} A^2b & Ab & b \end{bmatrix}\begin{bmatrix} 1 & 0 & 0 \\ \alpha_2 & 1 & 0 \\ \alpha_1 & \alpha_2 & 1 \end{bmatrix},$$

$$= \begin{bmatrix} 0 & 0 & 1 \end{bmatrix}\begin{bmatrix} 16 & 4 & 2 \\ 8 & 6 & 1 \\ 12 & 2 & 1 \end{bmatrix}\begin{bmatrix} 1 & 0 & 0 \\ 0 & 1 & 0 \\ -9 & 0 & 1 \end{bmatrix} = \begin{bmatrix} 3 & 2 & 1 \end{bmatrix}.$$

So the controllability canonical form I of the system is:

$$\dot{\bar{x}} = \begin{bmatrix} 0 & 1 & 0 \\ 0 & 0 & 1 \\ -2 & 9 & 0 \end{bmatrix}\bar{x} + \begin{bmatrix} 0 \\ 0 \\ 1 \end{bmatrix}u,$$

$$y = \begin{bmatrix} 3 & 2 & 1 \end{bmatrix}\bar{x}.$$

From equation (5.56) we can write out the transfer function of the system as:

$$W(s) = \frac{\beta_2 s^2 + \beta_1 s + \beta_0}{s^3 + \alpha_2 s^2 + \alpha_1 s + \alpha_0} = \frac{s^2 + 2s + 3}{s^3 - 9s + 2}.$$

Example 5.13. Try to transform the state space equation of Example 5.11 into controllability canonical form II.

Solution. As shown in Example 5.11, we know $\alpha_2 = 0$, $\alpha_1 = -9$, $\alpha_0 = 2$. From (5.60) to (5.61), we have:

$$\bar{A} = \begin{bmatrix} 0 & 0 & -\alpha_0 \\ 1 & 0 & -\alpha_1 \\ 0 & 1 & -\alpha_2 \end{bmatrix} = \begin{bmatrix} 0 & 0 & -2 \\ 1 & 0 & 9 \\ 0 & 1 & 0 \end{bmatrix},$$

$$\bar{b} = \begin{bmatrix} 1 \\ 0 \\ 0 \end{bmatrix},$$

$$\bar{C} = \begin{bmatrix} Cb & CAb & CA^2b \end{bmatrix} = \begin{bmatrix} 1 & 2 & 12 \end{bmatrix}.$$

Thus, the controllability canonical form II of the system is:

$$\dot{\bar{x}} = \begin{bmatrix} 0 & 0 & -2 \\ 1 & 0 & 9 \\ 0 & 1 & 0 \end{bmatrix}\bar{x} + \begin{bmatrix} 1 \\ 0 \\ 0 \end{bmatrix}u,$$

$$y = \begin{bmatrix} 1 & 2 & 12 \end{bmatrix}\bar{x}.$$

Example 5.14. Try to transform the state space equation of Example 5.11 into observability canonical forms.

Solution. Step 1: Find the observability canonical form I of the system.
First we compute the observability matrix Q_o:

$$Q_o = \begin{bmatrix} C \\ CA \\ CA^2 \end{bmatrix} = \begin{bmatrix} 0 & 0 & 1 \\ 0 & 2 & 0 \\ 6 & -2 & 2 \end{bmatrix}.$$

rank $Q_o = 3$, so the system can be transformed into observability canonical forms. From equations (5.70) and (5.71), we have:

$$\tilde{A} = \begin{bmatrix} 0 & 1 & 0 \\ 0 & 0 & 1 \\ -2 & 9 & 0 \end{bmatrix}, \quad \tilde{b} = \begin{bmatrix} 1 \\ 2 \\ 12 \end{bmatrix}, \quad \tilde{C} = \begin{bmatrix} 1 & 0 & 0 \end{bmatrix}.$$

Thus, the observability canonical form I of the system is:

$$\dot{\tilde{x}} = \begin{bmatrix} 0 & 1 & 0 \\ 0 & 0 & 1 \\ -2 & 9 & 0 \end{bmatrix} \tilde{x} + \begin{bmatrix} 1 \\ 2 \\ 12 \end{bmatrix} u,$$

$$y = \begin{bmatrix} 1 & 0 & 0 \end{bmatrix} \tilde{x}.$$

Step 2: Find the observability canonical form II of the system.
From equations (5.76) and (5.77), we have:

$$\tilde{A} = \begin{bmatrix} 0 & 0 & -2 \\ 1 & 0 & 9 \\ 0 & 1 & 0 \end{bmatrix}, \quad \tilde{b} = \begin{bmatrix} 3 \\ 2 \\ 1 \end{bmatrix}, \quad \tilde{C} = \begin{bmatrix} 0 & 0 & 1 \end{bmatrix}.$$

Thus, the observability canonical form II of the system is:

$$\dot{\tilde{x}} = \begin{bmatrix} 0 & 0 & -2 \\ 1 & 0 & 9 \\ 0 & 1 & 0 \end{bmatrix} \tilde{x} + \begin{bmatrix} 3 \\ 2 \\ 1 \end{bmatrix} u,$$

$$y = \begin{bmatrix} 0 & 0 & 1 \end{bmatrix} \tilde{x}.$$

5.5.4 Observability and Controllability Canonical Form of Multiple Input Multiple Output Systems

Consider the following transfer function:

$$G(s) = \frac{B_{n-1}s^{n-1} + \cdots + B_1 s + B_0}{s^n + a_{n-1}s^{n-1} + \cdots + a_0} + B_n, \tag{5.79}$$

where B_i is $m \times r$ dimensional matrix.

The *controllable canonical form* is:

$$\dot{x} = \begin{bmatrix} O_r & I_r & & & \\ O_r & & \ddots & & \\ \vdots & & & \ddots & \\ O_r & & & & I_r \\ -a_0 I_r & & & & -a_{n-1}I_r \end{bmatrix} x + \begin{bmatrix} O_r \\ O_r \\ \vdots \\ O_r \\ I_r \end{bmatrix} u \qquad (5.80)$$

$$y = \begin{bmatrix} B_0 & B_1 & \cdots & B_{n-1} \end{bmatrix} x + B_n u ,$$

where O_r and I_r are $r \times r$ dimensional zero matrix and unit matrix, while r stands for dimension of input vector and n is the order of the denominator polynomial.

Example 5.15. The system transfer function matrix is given below. Try to give a controllable canonical form of the system:

$$W(s) = \begin{bmatrix} \frac{s+2}{s+1} & \frac{1}{s+3} \\ \frac{s}{s+1} & \frac{s+1}{s+2} \end{bmatrix} = \begin{bmatrix} \frac{1}{s+1} & \frac{1}{s+3} \\ -\frac{1}{s+1} & -\frac{1}{s+2} \end{bmatrix} + \begin{bmatrix} 1 & 0 \\ 1 & 1 \end{bmatrix}$$

$$= \frac{\begin{bmatrix} s^2+5s+6 & s^2+3s+2 \\ -(s^2+5s+6) & -(s^2+4s+3) \end{bmatrix}}{(s+1)(s+2)(s+3)} + \begin{bmatrix} 1 & 0 \\ 1 & 1 \end{bmatrix}$$

$$= \frac{\begin{bmatrix} 1 & 1 \\ -1 & -1 \end{bmatrix}s^2 + \begin{bmatrix} 5 & 3 \\ -5 & -4 \end{bmatrix}s + \begin{bmatrix} 6 & 2 \\ -6 & -3 \end{bmatrix}}{s^3+6s^2+11s+6} + \begin{bmatrix} 1 & 0 \\ 1 & 1 \end{bmatrix}.$$

Solution. For the system shown with the above model, we have $n = 3$, $m = r = 2$ and:

$$a_0 = 6, \quad a_1 = 11, \quad a_2 = 6,$$

$$D = \begin{bmatrix} 1 & 0 \\ 1 & 1 \end{bmatrix}, \quad B_0 = \begin{bmatrix} 6 & 2 \\ -6 & -3 \end{bmatrix}, \quad B_1 = \begin{bmatrix} 5 & 3 \\ -5 & -4 \end{bmatrix}, \quad B_2 = \begin{bmatrix} 1 & 1 \\ -1 & -1 \end{bmatrix}.$$

So we get the matrix of the controllable canonical form as follows:

$$A = \begin{bmatrix} 0 & 0 & 1 & 0 & 0 & 0 \\ 0 & 0 & 0 & 1 & 0 & 0 \\ 0 & 0 & 0 & 0 & 1 & 0 \\ 0 & 0 & 0 & 0 & 0 & 1 \\ -6 & 0 & -11 & 0 & -6 & 0 \\ 0 & -6 & 0 & -11 & 0 & -6 \end{bmatrix}, \quad B = \begin{bmatrix} 0 & 0 \\ 0 & 0 \\ 0 & 0 \\ 0 & 0 \\ 1 & 0 \\ 0 & 1 \end{bmatrix},$$

$$C = \begin{bmatrix} 6 & 2 & 5 & 3 & 1 & 1 \\ -6 & -3 & -5 & -4 & -1 & -1 \end{bmatrix}.$$

The *observable canonical form* is:

$$\dot{x} = \begin{bmatrix} 0_m & 0_m & \cdots & 0_m & 0_m & -\alpha_0 I_m \\ I_m & 0_m & \cdots & 0_m & 0_m & -\alpha_1 I_m \\ 0_m & I_m & \cdots & 0_m & 0_m & -\alpha_2 I_m \\ \vdots & \vdots & \cdots & \vdots & \vdots & \vdots \\ 0_m & 0_m & \cdots & 0_m & I_m & -\alpha_{n-1} I_m \end{bmatrix} x + \begin{bmatrix} B_0 \\ B_1 \\ \vdots \\ B_{n-1} \end{bmatrix} u$$

$$y = \begin{bmatrix} 0_m & 0_m & \cdots & I_m \end{bmatrix} x + B_n u \, ,$$

(5.81)

where 0_m and I_m are $m \times m$ dimensional zero matrix and unit matrix, while m stands for dimension of output vector, and n is the order of the denominator polynomial.

5.6 System Decomposition

5.6.1 Controllability Decomposition

Suppose the LTI system

$$\dot{x} = Ax + Bu$$
$$y = Cx$$

(5.82)

is partly controllable. The controllability matrix is:

$$Q_c = \begin{bmatrix} B & AB & \cdots & A^{n-1}B \end{bmatrix} \, ,$$

and:

$$\text{rank } Q_c = n_1 < n \, .$$

Therefore, there exists a nonsingular transformation of:

$$x = R_c \widehat{x} \, ,$$

(5.83)

Which can transfer the state equation into the following form:

$$\dot{\widehat{x}} = \widehat{A}\widehat{x} + \widehat{B}u$$
$$y = \widehat{C}\widehat{x} \, ,$$

(5.84)

where:

$$\widehat{x} = \begin{bmatrix} \widehat{x}_1 \\ \hline \widehat{x}_2 \end{bmatrix} \begin{matrix} n_1 \\ n - n_1 \end{matrix} \, ,$$

$$\widehat{A} = R_c^{-1} A R_c = \begin{bmatrix} \widehat{A}_{11} & \vdots & \widehat{A}_{12} \\ \hline 0 & \vdots & \widehat{A}_{22} \end{bmatrix} \begin{matrix} n_1 \\ n - n_1 \end{matrix} \, ,$$

(5.85)

$$\begin{matrix} n_1 & \vdots & n - n_1 \end{matrix}$$

$$\widehat{B} = R_c^{-1} B = \left[\begin{array}{c} \widehat{B}_1 \\ \hline 0 \end{array} \right] \begin{array}{c} n_1 \\ n - n_1 \end{array}, \tag{5.86}$$

$$\widehat{C} = C R_c = \left[\begin{array}{c|c} \widehat{C}_1 & \widehat{C}_2 \\ \hline n_1 & n - n_1 \end{array} \right]. \tag{5.87}$$

From the above equations, we know that after the system is transformed into equation (5.84), the state space description of the system is decomposed into controllable and uncontrollable parts. The n_1-dimensional subspace $\dot{\widehat{x}}_1 = \widehat{A}_{11}\widehat{x}_1 + \widehat{B}_1 u + \widehat{A}_{12}\widehat{x}_2$ is controllable, while the $(n - n_1)$-dimensional subspace $\dot{\widehat{x}}_2 = \widehat{A}_{22}\widehat{x}_2$ is uncontrollable. The state decomposition is shown in Figure 5.5. Because $\widehat{x}_2$ cannot be detected from u, $\widehat{x}_2$ only has uncontrollable free response. Obviously, if we neglect the $(n - n_1)$-dimensional subsystem, we can obtain a controllable system with less dimension.

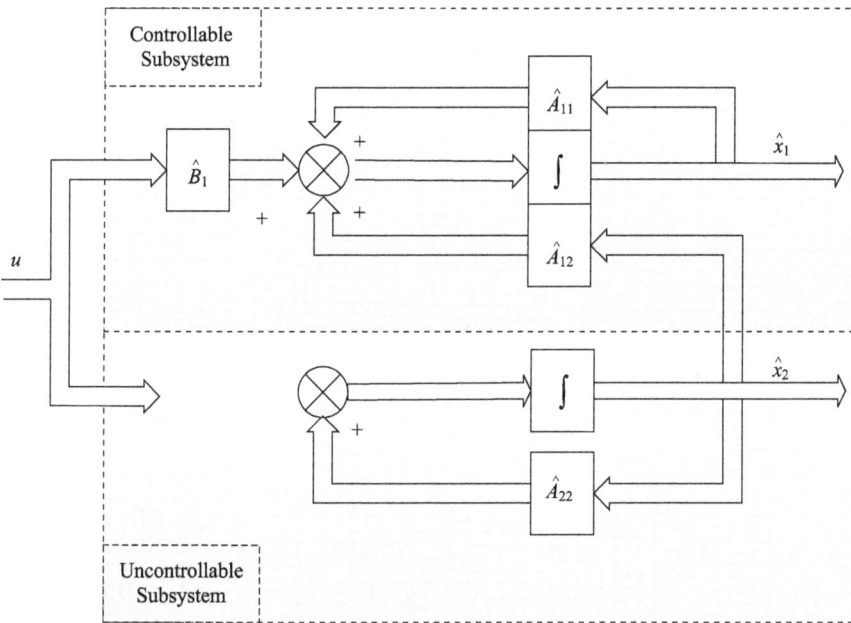

Fig. 5.5: Controllability decomposition of system.

We form the nonsingular transfer matrix:

$$R_c = \begin{bmatrix} R_1 & R_2 & \cdots & R_{n_1} & \cdots & R_n \end{bmatrix}, \tag{5.88}$$

where the first n_1 columns are any linearly independent columns of controllability matrix Q_c, and the remaining columns can be chosen as long as R_c is nonsingular.

Example 5.16. Are the state equations

$$\dot{x} = \begin{bmatrix} 0 & 0 & -1 \\ 1 & 0 & -3 \\ 0 & 1 & -3 \end{bmatrix} x + \begin{bmatrix} 1 \\ 1 \\ 0 \end{bmatrix} u$$

$$y = \begin{bmatrix} 0 & 1 & -2 \end{bmatrix} x$$

controllable? If not, please write the controllability decomposition of the system.

Solution. The controllability matrix is:

$$Q_c = \begin{bmatrix} b & Ab & A^2b \end{bmatrix} = \begin{bmatrix} 1 & 0 & -1 \\ 1 & 1 & -3 \\ 0 & 1 & -2 \end{bmatrix},$$

and rank $Q_c = 2 < n$. Therefore, the system is partly controllable.
We form the nonsingular transfer matrix as equation (5.88).

$$R_1 = b = \begin{bmatrix} 1 \\ 1 \\ 0 \end{bmatrix}, \quad R_2 = Ab = \begin{bmatrix} 0 \\ 1 \\ 1 \end{bmatrix}, \quad R_3 = \begin{bmatrix} 0 \\ 0 \\ 1 \end{bmatrix},$$

Thus:

$$R_c = \begin{bmatrix} 1 & 0 & 0 \\ 1 & 1 & 0 \\ 0 & 1 & 1 \end{bmatrix},$$

where R_3 is chosen arbitrarily as long as R_c is nonsingular.
After the transformation, the new state equation is:

$$\dot{\hat{x}} = R_c^{-1} A R_c \hat{x} + R_c^{-1} b u$$

$$= \begin{bmatrix} 1 & 0 & 0 \\ 1 & 1 & 0 \\ 0 & 1 & 1 \end{bmatrix}^{-1} \begin{bmatrix} 0 & 0 & -1 \\ 1 & 0 & -3 \\ 0 & 1 & -3 \end{bmatrix} \begin{bmatrix} 1 & 0 & 0 \\ 1 & 1 & 0 \\ 0 & 1 & 1 \end{bmatrix} \hat{x} + \begin{bmatrix} 1 & 0 & 0 \\ 1 & 1 & 0 \\ 0 & 1 & 1 \end{bmatrix}^{-1} \begin{bmatrix} 1 \\ 1 \\ 0 \end{bmatrix} u$$

$$= \begin{bmatrix} 0 & -1 & -1 \\ 1 & -2 & -2 \\ \hline 0 & 0 & -1 \end{bmatrix} \hat{x} + \begin{bmatrix} 1 \\ 0 \\ 0 \end{bmatrix} u,$$

$$y = C R_c \hat{x} = \begin{bmatrix} 1 & -1 & -2 \end{bmatrix} \hat{x}.$$

We choose:

$$R_3 = \begin{bmatrix} 1 \\ 0 \\ 1 \end{bmatrix}, \quad R_c = \begin{bmatrix} 1 & 0 & 1 \\ 1 & 1 & 0 \\ 0 & 1 & 1 \end{bmatrix}.$$

Thus:

$$\dot{\hat{x}} = \left[\begin{array}{cc|c} 0 & -1 & 0 \\ 1 & -2 & -2 \\ \hline 0 & 0 & -1 \end{array}\right] \hat{x} + \left[\begin{array}{c} 1 \\ 0 \\ \hline 0 \end{array}\right] u \, ,$$

$$y = CR_c\hat{x} = \begin{bmatrix} 1 & -1 & -2 \end{bmatrix} \hat{x} \, .$$

5.6.2 Observability Decomposition

Suppose the LTI system

$$\dot{x} = Ax + Bu$$
$$y = Cx$$

(5.89)

is partly observable. The observability matrix is:

$$Q_o = \begin{bmatrix} C \\ CA \\ \vdots \\ CA^{n-1} \end{bmatrix} \, ,$$

and:

$$\text{rank}\, Q_o = n_1 < n \, .$$

Therefore, there exists a nonsingular transformation:

$$x = R_o\tilde{x} \, ,$$

(5.90)

which can transfer the state equation into the following form:

$$\dot{\tilde{x}} = \tilde{A}\tilde{x} + \tilde{B}u$$
$$y = \tilde{C}\tilde{x} \, ,$$

(5.91)

where:

$$\tilde{x} = \left[\begin{array}{c} \tilde{x}_1 \\ \hline \tilde{x}_2 \end{array}\right] \begin{array}{l} n_1 \\ n - n_1 \end{array} \, ,$$

$$\tilde{A} = R_o^{-1}AR_o = \left[\begin{array}{c|c} \tilde{A}_{11} & 0 \\ \hline \tilde{A}_{21} & \tilde{A}_{22} \end{array}\right] \begin{array}{l} n_1 \\ n - n_1 \end{array} \, ,$$

$$\begin{array}{cc} n_1 & n - n_1 \end{array}$$

(5.92)

$$\tilde{B} = R_o^{-1}B = \left[\begin{array}{c} \tilde{B}_1 \\ \hline \tilde{B}_2 \end{array}\right] \begin{array}{l} n_1 \\ n - n_1 \end{array} \, ,$$

(5.93)

$$\tilde{C} = CR_o = \begin{array}{c} \left[\begin{array}{c|c} \tilde{C}_1 & 0 \end{array}\right] \\ \begin{array}{cc} n_1 & n - n_1 \end{array} \end{array} \, .$$

(5.94)

From the above equations, we know that when the system is transferred into equation (5.91), the state space description of the system is decomposed into observable and unobservable parts. The n_1-dimensional subspace

$$\dot{\tilde{x}}_1 = \tilde{A}_{11}\tilde{x}_1 + \tilde{B}_1 u$$
$$y = \tilde{C}_1 \tilde{x}_1$$

is observable, while the $(n - n_1)$-dimensional subspace

$$\dot{\tilde{x}}_2 = \tilde{A}_{21}\tilde{x}_1 + \tilde{A}_{22}\tilde{x}_2 + \tilde{B}_2 u$$

is unobservable. The state decomposition is shown in Figure 5.6. Obviously, if we do not consider the $(n-n_1)$-dimensional unobservable subsystem, we can obtain a n_1-dimensional observable system.

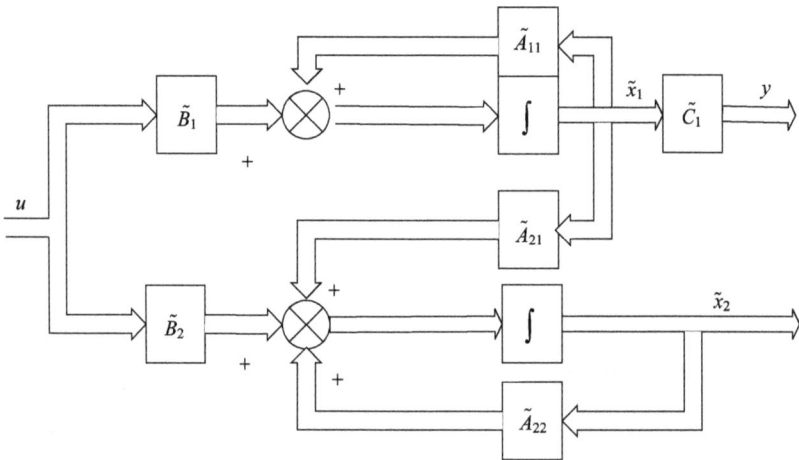

Fig. 5.6: Observability decomposition of a system.

We form the nonsingular transfer matrix:

$$R_0^{-1} = \begin{bmatrix} R_1' \\ R_2' \\ \vdots \\ R_{n_1}' \\ \vdots \\ R_n' \end{bmatrix}, \tag{5.95}$$

where the first n_1 rows are any linearly independent rows of observability matrix Q_0, and the remaining rows can be chosen as long as R_0^{-1} is nonsingular.

Example 5.17. Are the state equations

$$\dot{x} = \begin{bmatrix} 0 & 0 & -1 \\ 1 & 0 & -3 \\ 0 & 1 & -3 \end{bmatrix} x + \begin{bmatrix} 1 \\ 1 \\ 0 \end{bmatrix} u$$

$$y = \begin{bmatrix} 0 & 1 & -2 \end{bmatrix} x$$

observable? If not, please write the observability decomposition of the system.

Solution. The observability matrix is:

$$Q_o = \begin{bmatrix} C \\ CA \\ CA^2 \end{bmatrix} = \begin{bmatrix} 0 & 1 & -2 \\ 1 & -2 & 3 \\ -2 & 3 & -4 \end{bmatrix},$$

and rank $Q_o = 2 < n$. Therefore, the system is partly observable.
We form the nonsingular transfer matrix as equation (5.95).

$$R_1' = C = \begin{bmatrix} 0 & 1 & -2 \end{bmatrix}, \quad R_2' = CA = \begin{bmatrix} 1 & -2 & 3 \end{bmatrix}, \quad R_3' = \begin{bmatrix} 0 & 0 & 1 \end{bmatrix},$$

thus,

$$R_o^{-1} = \begin{bmatrix} 0 & 1 & -2 \\ 1 & -2 & 3 \\ 0 & 0 & 1 \end{bmatrix}$$

and

$$R_o = \begin{bmatrix} 2 & 1 & 1 \\ 1 & 0 & 2 \\ 0 & 0 & 1 \end{bmatrix}$$

where R_3 is chosen arbitrarily as long as R_o^{-1} is nonsingular.
After the transformation, the new state equation is:

$$\dot{\tilde{x}} = R_o^{-1} A R_o \tilde{x} + R_o^{-1} bu$$

$$= \begin{bmatrix} 0 & -1 & 0 \\ -1 & -2 & 0 \\ 1 & 0 & -1 \end{bmatrix} \tilde{x} + \begin{bmatrix} 1 \\ -1 \\ 0 \end{bmatrix} u,$$

$$y = CR_o\tilde{x} = \begin{bmatrix} 1 & 0 & 0 \end{bmatrix} \tilde{x}.$$

5.6.3 Controllability and Observability Decomposition

Suppose the LTI system

$$\dot{x} = Ax + Bu$$
$$y = Cx \tag{5.96}$$

is partly controllable and observable. Therefore, there exists a nonsingular transformation:

$$x = R\bar{x}, \tag{5.97}$$

which can transfer the state equation into the following form:

$$\dot{\bar{x}} = \bar{A}\bar{x} + \bar{B}v$$
$$y = \bar{C}\bar{x}. \tag{5.98}$$

Here:

$$\bar{A} = R^{-1}AR = \begin{bmatrix} A_{11} & 0 & A_{13} & 0 \\ A_{21} & A_{22} & A_{23} & A_{24} \\ 0 & 0 & A_{33} & 0 \\ 0 & 0 & A_{43} & A_{44} \end{bmatrix}, \tag{5.99}$$

$$\bar{B} = R^{-1}B = \begin{bmatrix} B_1 \\ B_2 \\ 0 \\ 0 \end{bmatrix}, \tag{5.100}$$

$$\bar{C} = CR = \begin{bmatrix} C_1 & 0 & C_3 & 0 \end{bmatrix}. \tag{5.101}$$

From the configuration of $\bar{A}, \bar{B}, \bar{C}$, we know that the n-dimensional state space is divided into four subspaces according to the controllability and observability of the system. Equation (5.98) can be rewritten as follows:

$$\begin{bmatrix} \dot{x}_{co} \\ \dot{x}_{c\bar{o}} \\ \dot{x}_{\bar{c}o} \\ \dot{x}_{\bar{c}\bar{o}} \end{bmatrix} = \begin{bmatrix} A_{11} & 0 & A_{13} & 0 \\ A_{21} & A_{22} & A_{23} & A_{24} \\ 0 & 0 & A_{33} & 0 \\ 0 & 0 & A_{43} & A_{44} \end{bmatrix} \begin{bmatrix} x_{co} \\ x_{c\bar{o}} \\ x_{\bar{c}o} \\ x_{\bar{c}\bar{o}} \end{bmatrix} + \begin{bmatrix} B_1 \\ B_2 \\ 0 \\ 0 \end{bmatrix} u$$

$$y = \begin{bmatrix} C_1 & 0 & C_3 & 0 \end{bmatrix} \begin{bmatrix} x_{co} \\ x_{c\bar{o}} \\ x_{\bar{c}o} \\ x_{\bar{c}\bar{o}} \end{bmatrix}, \tag{5.102}$$

and the subsystem (A_{11}, B_1, C_1) is controllable and observable.

The block diagram of equation (5.98) is shown in Figure 5.7.

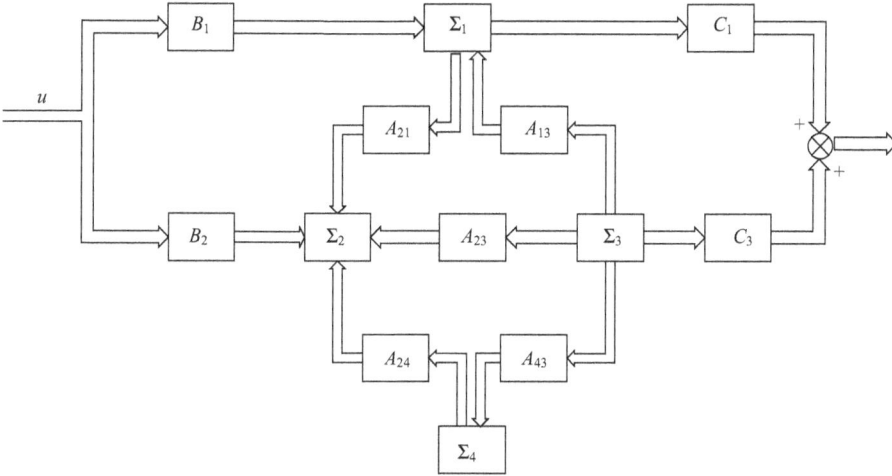

Fig. 5.7: The block diagram of equation (5.98).

From Figure 5.7, we know the condition for transfer of the information between the four subsystems.

1. The steps of transposition

Step 1. The transfer matrix

$$x = R_c \begin{bmatrix} x_c \\ x_{\bar{c}} \end{bmatrix} \tag{5.103}$$

will transform system $\Sigma = (A, B, C)$ into:

$$\begin{aligned} \begin{bmatrix} \dot{x}_c \\ \dot{x}_{\bar{c}} \end{bmatrix} &= R_c^{-1} A R_c \begin{bmatrix} x_c \\ x_{\bar{c}} \end{bmatrix} + R_c^{-1} B u \\ &= \begin{bmatrix} \bar{A}_1 & \bar{A}_2 \\ 0 & \bar{A}_4 \end{bmatrix} \begin{bmatrix} x_c \\ x_{\bar{c}} \end{bmatrix} + \begin{bmatrix} \bar{B} \\ 0 \end{bmatrix} u \,, \end{aligned} \tag{5.104}$$

$$y = C R_c \begin{bmatrix} x_c \\ x_{\bar{c}} \end{bmatrix} = \begin{bmatrix} \bar{C}_1 & \bar{C}_2 \end{bmatrix} \begin{bmatrix} x_c \\ x_{\bar{c}} \end{bmatrix} \,,$$

where x_c is the controllable state, $x_{\bar{c}}$ is the uncontrollable state, and R_c is formed according to equation (5.88).

Step 2. The transfer matrix $x_{\bar{c}} = R_{o2} [x_{\bar{c}o} \ x_{\overline{co}}]^T$ will transform the uncontrollable subsystem $\Sigma_{\bar{c}} = (\bar{A}_4, 0, \bar{C}_2)$ into:

$$\begin{bmatrix} \dot{x}_{\bar{c}o} \\ \dot{x}_{\overline{co}} \end{bmatrix} = R_{o2}^{-1} \bar{A}_4 R_{o2} \begin{bmatrix} x_{\bar{c}o} \\ x_{\overline{co}} \end{bmatrix} = \begin{bmatrix} A_{33} & 0 \\ A_{43} & A_{44} \end{bmatrix} \begin{bmatrix} x_{\bar{c}o} \\ x_{\overline{co}} \end{bmatrix} \,,$$

$$y_2 = \bar{C}_2 R_{o2} \begin{bmatrix} x_{\bar{c}o} \\ x_{\overline{co}} \end{bmatrix} = \begin{bmatrix} C_3 & 0 \end{bmatrix} \begin{bmatrix} x_{\bar{c}o} \\ x_{\overline{co}} \end{bmatrix} \,,$$

where $x_{\bar{c}o}$ is an uncontrollable but observable state, $x_{\overline{co}}$ is an uncontrollable and unobservable state, and R_{o2} is formed according to equation (5.95) for system $\Sigma_{\bar{c}} = (\overline{A}_4, 0, \overline{C}_2)$.

Step 3. The transformation $x_c = R_{o1}[x_{co}\ x_{c\bar{o}}]^T$ will transform the controllable subsystem $\Sigma_c = (\overline{A}_1, B, \overline{C}_1)$ according to observability.

According to equation (5.104), we can obtain:

$$\dot{x}_c = \overline{A}_1 x_c + \overline{A}_2 x_{\bar{c}} + Bu .$$

Substitute the state transfer equations into the above equation:

$$R_{o1}\begin{bmatrix} \dot{x}_{co} \\ \dot{x}_{c\bar{o}} \end{bmatrix} = \overline{A}_1 R_{o1}\begin{bmatrix} x_{co} \\ x_{c\bar{o}} \end{bmatrix} + \overline{A}_2 R_{o2}\begin{bmatrix} x_{\bar{c}o} \\ x_{\overline{co}} \end{bmatrix} + Bu .$$

Multiply the above equation by R_{o1}^{-1}, and we will have:

$$\begin{bmatrix} \dot{x}_{co} \\ \dot{x}_{c\bar{o}} \end{bmatrix} = R_{o1}^{-1}\overline{A}_1 R_{o1}\begin{bmatrix} x_{co} \\ x_{c\bar{o}} \end{bmatrix} + R_{o1}^{-1}\overline{A}_2 R_{o2}\begin{bmatrix} x_{\bar{c}o} \\ x_{\overline{co}} \end{bmatrix} + R_{o1}^{-1}Bu$$

$$= \begin{bmatrix} A_{11} & 0 \\ A_{21} & A_{22} \end{bmatrix}\begin{bmatrix} x_{co} \\ x_{c\bar{o}} \end{bmatrix} + \begin{bmatrix} A_{13} & 0 \\ A_{23} & A_{24} \end{bmatrix}\begin{bmatrix} x_{\bar{c}o} \\ x_{\overline{co}} \end{bmatrix} + \begin{bmatrix} B_1 \\ B_2 \end{bmatrix}u ,$$

$$y = \overline{C}R_{o1}\begin{bmatrix} x_{co} \\ x_{c\bar{o}} \end{bmatrix} = \begin{bmatrix} C_1 & 0 \end{bmatrix}\begin{bmatrix} x_{co} \\ x_{c\bar{o}} \end{bmatrix} ,$$

where x_{co} is controllable and observable state, $x_{c\bar{o}}$ is controllable but observable state, and R_{o1} is formed according to equation (5.95) for system $\Sigma_c = (\overline{A}_1, B, \overline{C}_1)$.

After the above three transformations, we can have the decomposition description as follows, according to controllability and observability:

$$\begin{bmatrix} \dot{x}_{co} \\ \dot{x}_{c\bar{o}} \\ \dot{x}_{\bar{c}o} \\ \dot{x}_{\overline{co}} \end{bmatrix} = \begin{bmatrix} A_{11} & 0 & A_{13} & 0 \\ A_{21} & A_{22} & A_{23} & A_{24} \\ 0 & 0 & A_{33} & 0 \\ 0 & 0 & A_{43} & A_{44} \end{bmatrix}\begin{bmatrix} x_{co} \\ x_{c\bar{o}} \\ x_{\bar{c}o} \\ x_{\overline{co}} \end{bmatrix} + \begin{bmatrix} B_1 \\ B_2 \\ 0 \\ 0 \end{bmatrix}u$$

$$y = \begin{bmatrix} C_1 & 0 & C_3 & 0 \end{bmatrix}\begin{bmatrix} x_{co} \\ x_{c\bar{o}} \\ x_{\bar{c}o} \\ x_{\overline{co}} \end{bmatrix} .$$

Example 5.18. The LTI system

$$\dot{x} = \begin{bmatrix} 0 & 0 & -1 \\ 1 & 0 & -3 \\ 0 & 1 & -3 \end{bmatrix}x + \begin{bmatrix} 1 \\ 1 \\ 0 \end{bmatrix}u$$

$$y = \begin{bmatrix} 0 & 1 & -3 \end{bmatrix}x$$

is partly controllable and observable. Give the controllability and observability decomposition.

Solution.

$$R_c = \begin{bmatrix} 1 & 0 & 0 \\ 1 & 1 & 0 \\ 0 & 1 & 1 \end{bmatrix}.$$

After transformation, we have:

$$\begin{bmatrix} \dot{x}_c \\ \dot{x}_{\bar{c}} \end{bmatrix} = \begin{bmatrix} 0 & -1 & -1 \\ 1 & -2 & -2 \\ \hline 1 & 0 & -1 \end{bmatrix} \begin{bmatrix} x_c \\ x_{\bar{c}} \end{bmatrix} + \begin{bmatrix} 1 \\ 0 \\ \hline 0 \end{bmatrix} u,$$

$$y = \begin{bmatrix} 1 & -1 & 2 \end{bmatrix} \begin{bmatrix} x_c \\ x_{\bar{c}} \end{bmatrix}.$$

From the above, we know that the uncontrollable subspace $x_{\bar{c}}$ is one-dimensional. Obviously, the subspace is observable. Therefore, there is no need to decompose the subspace.

The controllable subsystem Σ_c is:

$$\dot{x}_c = \begin{bmatrix} 0 & -1 \\ 1 & -2 \end{bmatrix} x_c + \begin{bmatrix} -1 \\ -2 \end{bmatrix} x_{\bar{c}} + \begin{bmatrix} 1 \\ 0 \end{bmatrix} u,$$

$$y_1 = \begin{bmatrix} 1 & -1 \end{bmatrix} x_c.$$

Then, we decompose the subsystem Σ_c according to observability.

We form the nonsingular matrix according to equation (5.95):

$$R_o^{-1} = \begin{bmatrix} 1 & -1 \\ 0 & 1 \end{bmatrix},$$

which transfers the system Σ_c into:

$$\begin{bmatrix} \dot{x}_{co} \\ \dot{x}_{c\bar{o}} \end{bmatrix} = \begin{bmatrix} 1 & -1 \\ 0 & 1 \end{bmatrix} \begin{bmatrix} 0 & -1 \\ 1 & -2 \end{bmatrix} \begin{bmatrix} 1 & -1 \\ 0 & 1 \end{bmatrix}^{-1} \begin{bmatrix} x_{co} \\ x_{c\bar{o}} \end{bmatrix}$$

$$+ \begin{bmatrix} 1 & -1 \\ 0 & 1 \end{bmatrix} \begin{bmatrix} -1 \\ -2 \end{bmatrix} x_{\bar{c}} + \begin{bmatrix} 1 & -1 \\ 0 & 1 \end{bmatrix}^{-1} \begin{bmatrix} 1 \\ 0 \end{bmatrix} u.$$

Equivalently:

$$\begin{bmatrix} \dot{x}_{co} \\ \dot{x}_{c\bar{o}} \end{bmatrix} = \begin{bmatrix} -1 & 0 \\ 1 & -1 \end{bmatrix} \begin{bmatrix} x_{co} \\ x_{c\bar{o}} \end{bmatrix} + \begin{bmatrix} 1 \\ -2 \end{bmatrix} x_{\bar{c}} + \begin{bmatrix} 1 \\ 0 \end{bmatrix} u,$$

$$y_1 = \begin{bmatrix} 1 & -1 \end{bmatrix} \begin{bmatrix} 1 & -1 \\ 0 & 1 \end{bmatrix}^{-1} \begin{bmatrix} x_{co} \\ x_{c\bar{o}} \end{bmatrix} = \begin{bmatrix} 1 & 0 \end{bmatrix} \begin{bmatrix} x_{co} \\ x_{c\bar{o}} \end{bmatrix}.$$

With the above two transformations, we have the decomposition description:

$$\begin{bmatrix} \dot{x}_{co} \\ \dot{x}_{c\bar{o}} \\ \dot{x}_{\bar{c}o} \end{bmatrix} = \begin{bmatrix} -1 & 0 & 1 \\ 1 & -1 & -2 \\ 0 & 0 & -1 \end{bmatrix} \begin{bmatrix} x_{co} \\ x_{c\bar{o}} \\ x_{\bar{c}o} \end{bmatrix} + \begin{bmatrix} 1 \\ 0 \\ 0 \end{bmatrix} u,$$

$$y = \begin{bmatrix} 1 & 0 & -2 \end{bmatrix} \begin{bmatrix} x_{co} \\ x_{c\bar{o}} \\ x_{\bar{c}o} \end{bmatrix}.$$

2. Alternative method of decomposition

First, we can transfer the system into Jordan canonical form, then examine the controllability and observability of all state variables according to controllability and observability criteria. Finally, we form the corresponding subsystems with those state variables.

For example, the given Jordan canonical form of system $\Sigma = (A, B, C)$ is:

$$\begin{bmatrix} \dot{x}_1 \\ \dot{x}_2 \\ \dot{x}_3 \\ \dot{x}_4 \\ \dot{x}_5 \\ \dot{x}_6 \end{bmatrix} = \begin{bmatrix} -4 & 1 & & & & \\ 0 & -4 & & & & \\ & & 3 & 1 & & \\ & & 0 & 3 & & \\ & & & & -1 & 1 \\ & & & & 0 & -1 \end{bmatrix} \begin{bmatrix} x_1 \\ x_2 \\ x_3 \\ x_4 \\ x_5 \\ x_6 \end{bmatrix} + \begin{bmatrix} 1 & 3 \\ 5 & 7 \\ 4 & 3 \\ 0 & 0 \\ 1 & 6 \\ 0 & 0 \end{bmatrix} \begin{bmatrix} u_1 \\ u_2 \end{bmatrix},$$

$$\begin{bmatrix} y_1 \\ y_2 \end{bmatrix} = \begin{bmatrix} 3 & 1 & 0 & 5 & 0 & 0 \\ 1 & 4 & 0 & 2 & 0 & 0 \end{bmatrix} \begin{bmatrix} x_1 \\ x_2 \\ x_3 \\ x_4 \\ x_5 \\ x_6 \end{bmatrix}.$$

According to the controllability and observability criteria of Jordan canonical forms, we know that x_1, x_2 are controllable and observable variables, x_3, x_5 are controllable but unobservable variables, x_4 are uncontrollable and observable variables and x_6 are uncontrollable and unobservable variables.

Thus:

$$x_{co} = \begin{bmatrix} x_1 \\ x_2 \end{bmatrix}, \quad x_{c\bar{o}} = \begin{bmatrix} x_3 \\ x_5 \end{bmatrix}, \quad x_{\bar{c}o} = x_4, \quad x_{\bar{c}\bar{o}} = x_6.$$

Rearrange according to this order, and you can get:

$$
\begin{bmatrix} \dot{x}_{co} \\ \dot{x}_{c\bar{o}} \\ \dot{x}_{\bar{c}o} \\ \dot{x}_{\bar{c}\bar{o}} \end{bmatrix} =
\left[\begin{array}{cc:cc:c:c}
-4 & 1 & 0 & 0 & 0 & 0 \\
0 & -4 & 0 & 0 & 0 & 0 \\
\hdashline
0 & 0 & 3 & 0 & 1 & 0 \\
0 & 0 & 0 & -1 & 0 & 1 \\
\hdashline
0 & 0 & 0 & 0 & 3 & 0 \\
\hdashline
0 & 0 & 0 & 0 & 0 & -1
\end{array}\right]
\begin{bmatrix} x_{co} \\ x_{c\bar{o}} \\ x_{\bar{c}o} \\ x_{\bar{c}\bar{o}} \end{bmatrix} +
\begin{bmatrix} 1 & 3 \\ 5 & 7 \\ \hdashline 4 & 3 \\ 1 & 6 \\ \hdashline 0 & 0 \\ \hdashline 0 & 0 \end{bmatrix}
\begin{bmatrix} u_1 \\ u_2 \end{bmatrix} ,
$$

$$\widetilde{A}_m = P^{-1} A_m P .$$

5.6.4 Minimum Realization

Definition of Realization
Given a transfer matrix $W(s)$, if a state space equation Σ exists, such as:

$$\dot{x} = Ax + Bu$$
$$y = Cx + Du ,$$

which has $W(s)$ as its transfer matrix, such as:

$$C(sI - A)^{-1}B + D = W(s) ,$$

then $W(s)$ is realizable and Σ is one of the realizations of $W(s)$.

It is noticeable that not every transfer matrix $W(s)$ is realizable. $W(s)$ is physically realizable if it meets the following criteria:
(i) If all the coefficients of the numerator and denominator polynomials of each element $W_{ik}(s)$ ($i = 1, 2, \ldots, m$; $k = 1, 2, \ldots, r$) are real constants.
(ii) If $W_{ik}(s)$ is a real rational fraction of s, i.e., the order of the numerator polynomial is no more than that of the denominator polynomial. When all the elements of $W(s)$ are strictly real rational fractions, the realization of $W(s)$ has the form of (A, B, C). Apart from this, the realization has the form of (A, B, C, D) and $D = \lim_{s \to 0} W(s)$.

Minimum Realization
If a transfer function is realizable, then it has an infinite number of realizations, although not necessarily of the same dimension. From the engineering point of view, it is significant to find the class of minimal dimensional realizations of the system.

1. Definition of minimum realization
Consider one realization of the transfer function $W(s)$, Σ:

$$\dot{x} = Ax + Bu$$
$$y = Cx .$$

If any other realization

$$\dot{\tilde{x}} = \widetilde{A}\tilde{x} + \widetilde{B}u$$
$$y = \widetilde{C}\tilde{x}$$

has more dimensions than Σ, then Σ is the minimum realization of the system.

Because the transfer matrix can only reflect the dynamic behaviors of the controllable and observable subsystem, removing the uncontrollable or unobservable states will not change the transfer matrix of the system. Thus, the state space expression with uncontrollable or unobservable states cannot be the minimum realization. As stated above, we have the following methods to verify the minimum realization.

2. Steps to find minimum realization

Theorem 5.12. *The realization of transfer matrix $W(s)$, Σ:*

$$\dot{x} = Ax + Bu$$
$$y = Cx,$$

is the minimum realization if, and only if, $\Sigma(A, B, C)$ is controllable and observable.

According to this theorem, we can find the minimum realization of any transfer matrix $W(s)$ whose elements are all strictly real rational fractions. Usually, we can obtain the minimum realization as follows:

(1) For a given transfer matrix $W(s)$, first we select one realization $\Sigma(A, B, C)$. More often, we choose the controllability canonical or observability canonical for the sake of convenience.

(2) For the $\Sigma(A, B, C)$ chosen above, we find its controllable and observable part $(\widetilde{A}_1, \widetilde{B}_1, \widetilde{C}_1)$. So this part is just the minimum realization of the system.

Example 5.19. Try to find the minimum realization of the transfer matrix:

$$W(s) = \left[\frac{1}{(s+1)(s+2)} \quad \frac{1}{(s+2)(s+3)} \right] .$$

Solution. $W(s)$ is a strictly real rational fraction of s. Rewrite it in the descending order of s in the following form:

$$W(s) = \left[\frac{s+3}{(s+1)(s+2)(s+3)} \quad \frac{s+1}{(s+1)(s+2)(s+3)} \right]$$

$$= \frac{1}{(s+1)(s+2)(s+3)} \left[(s+3) \quad (s+1) \right]$$

$$= \frac{1}{s^3 + 6s^2 + 11s + 6} \left\{ \begin{bmatrix} 1 & 1 \end{bmatrix} s + \begin{bmatrix} 3 & 1 \end{bmatrix} \right\} .$$

From equation (5.56), we know that:

$$\alpha_0 = 6, \qquad \alpha_1 = 11, \qquad \alpha_2 = 6$$
$$\beta_0 = \begin{bmatrix} 3 & 1 \end{bmatrix}, \qquad \beta_1 = \begin{bmatrix} 1 & 1 \end{bmatrix}, \qquad \beta_2 = \begin{bmatrix} 0 & 0 \end{bmatrix}.$$

The dimension of the output vector is $m = 1$, and the dimension of the input vector is $r = 2$. First we adopt the controllability canonical form realization:

$$A_0 = \begin{bmatrix} 0_m & 0_m & -\alpha_0 I_m \\ I_m & 0_m & -\alpha_1 I_m \\ 0_m & 0_m & -\alpha_2 I_m \end{bmatrix} = \begin{bmatrix} 0 & 0 & -6 \\ 1 & 0 & -11 \\ 0 & 1 & -6 \end{bmatrix}$$

$$B_0 = \begin{bmatrix} \beta_0 \\ \beta_1 \\ \beta_2 \end{bmatrix} = \begin{bmatrix} 3 & 1 \\ 1 & 1 \\ 0 & 0 \end{bmatrix}$$

$$C_0 = \begin{bmatrix} 0_m & 0_m & I_m \end{bmatrix} = \begin{bmatrix} 0 & 0 & 1 \end{bmatrix}$$

Then we check whether the realization $\Sigma = (A_0, B_0, C_0)$ is controllable or not:

$$Q_c = \begin{bmatrix} B_0 & A_0 B_0 & A_0^2 B_0 \end{bmatrix} = \begin{bmatrix} 3 & 1 & 0 & 0 & -6 & -6 \\ 1 & 1 & 3 & 1 & -11 & -11 \\ 0 & 0 & 1 & 1 & -3 & -5 \end{bmatrix},$$

rank $Q_c = 3 = n$.

Therefore, $\Sigma = (A_0, B_0, C_0)$ is controllable and observable, so it is the minimum realization.

Example 5.20. Try to find the minimum realization of the transfer matrix:

$$W(s) = \begin{bmatrix} \frac{s+2}{s+1} & \frac{1}{s+3} \\ \frac{s}{s+1} & \frac{s+1}{s+2} \end{bmatrix}$$

Solution. First, we simplify $W(s)$ into the form of strictly real rational function and write its controllability canonical form (or observability canonical form). After computing this, the controllability canonical form of the system is:

$$A = \begin{bmatrix} 0 & 0 & 1 & 0 & 0 & 0 \\ 0 & 0 & 0 & 1 & 0 & 0 \\ 0 & 0 & 0 & 0 & 1 & 0 \\ 0 & 0 & 0 & 0 & 0 & 1 \\ -6 & 0 & -11 & 0 & -6 & 0 \\ 0 & -6 & 0 & -4 & 0 & -6 \end{bmatrix}, \qquad B = \begin{bmatrix} 0 & 0 \\ 0 & 0 \\ 0 & 0 \\ 0 & 0 \\ 1 & 0 \\ 0 & 1 \end{bmatrix},$$

$$C = \begin{bmatrix} 6 & 2 & 5 & 3 & 1 & 1 \\ -6 & -3 & -5 & -4 & -1 & -1 \end{bmatrix}, \qquad D = \begin{bmatrix} 1 & 0 \\ 1 & 1 \end{bmatrix}.$$

We then examine whether the states realized by the controllability canonical form are observable or not:

$$
Q_0 = \begin{bmatrix} C \\ CA \\ CA^2 \end{bmatrix} = \begin{bmatrix}
6 & 2 & 5 & 3 & 1 & 1 \\
-6 & -3 & -5 & -4 & -1 & -1 \\
-6 & -6 & -5 & -9 & -1 & -3 \\
6 & 6 & 5 & 8 & 1 & 2 \\
6 & 18 & 5 & 27 & 1 & 9 \\
-6 & -12 & -5 & -16 & -1 & -4
\end{bmatrix}.
$$

As rank $Q_0 = 3 < n = 6$, the controllability canonical form is not the minimum realization. Thus, we decompose the structure according to the observability.

Now we construct the transfer matrix R_0^{-1} and decompose the system according to the observability:

$$
R_0^{-1} = \left[\begin{array}{ccc:ccc}
6 & 2 & 5 & 3 & 1 & 1 \\
-6 & -3 & -5 & -4 & -1 & -1 \\
-6 & -6 & -5 & -9 & -1 & -3 \\
\hdashline
1 & 0 & 0 & 0 & 0 & 0 \\
0 & 1 & 0 & 0 & 0 & 0 \\
0 & 0 & 1 & 0 & 0 & 0
\end{array}\right].
$$

Thus:

$$
R_0 = \left[\begin{array}{ccc:ccc}
0 & 0 & 0 & 1 & 0 & 0 \\
0 & 0 & 0 & 0 & 1 & 0 \\
0 & 0 & 0 & 0 & 0 & 1 \\
\hdashline
-1 & -1 & 0 & 0 & -1 & 0 \\
\frac{3}{2} & 0 & \frac{1}{2} & -6 & 0 & -5 \\
\frac{5}{2} & 3 & -\frac{1}{2} & 0 & 1 & 0
\end{array}\right].
$$

So:

$$
\widehat{A} = R_0^{-1} A R_0 = \left[\begin{array}{ccc:ccc}
0 & 0 & 1 & 0 & 0 & 0 \\
-\frac{3}{2} & -2 & -\frac{1}{2} & 0 & 0 & 0 \\
-3 & 0 & -4 & 0 & 0 & 0 \\
\hdashline
0 & 0 & 0 & 0 & 0 & 1 \\
-1 & -1 & 0 & 0 & -1 & 0 \\
\frac{3}{2} & 0 & -2 & -6 & 0 & -5
\end{array}\right] = \begin{bmatrix} \widehat{A}_{11} & 0 \\ \widehat{A}_{21} & \widehat{A}_{22} \end{bmatrix},
$$

$$
\widehat{B} = R_0^{-1} B = \left[\begin{array}{cc}
1 & 1 \\
-1 & -1 \\
-1 & -3 \\
\hdashline
0 & 0 \\
0 & 0 \\
0 & 0
\end{array}\right] = \begin{bmatrix} \widehat{B}_1 \\ 0 \end{bmatrix},
$$

$$\hat{C} = CR_0 = \begin{bmatrix} 1 & 0 & 0 & 0 & 0 & 0 \\ 0 & 1 & 0 & 0 & 0 & 0 \end{bmatrix} = \begin{bmatrix} \hat{C}_1 & 0 \end{bmatrix},$$

After examination, $\Sigma = (\hat{A}_{11}, \hat{B}_1, \hat{C}_1)$ is a controllable and observable subsystem. Thus, the minimum realization of $W(s)$ is:

$$A_m = \hat{A}_{11} = \begin{bmatrix} 0 & 0 & 1 \\ -\frac{3}{2} & -2 & -\frac{1}{2} \\ -3 & 0 & -4 \end{bmatrix}, \qquad B_m = \hat{B}_1 = \begin{bmatrix} 1 & 1 \\ -1 & -1 \\ -1 & -3 \end{bmatrix},$$

$$C_m = \hat{C}_1 = \begin{bmatrix} 1 & 0 & 0 \\ 0 & 1 & 0 \end{bmatrix}, \qquad D = \begin{bmatrix} 1 & 0 \\ 1 & 1 \end{bmatrix}.$$

After computing the transfer function according to the above A_m, B_m, C_m, D, we can check the result:

$$C_m(sI - A_m)^{-1}B_m + D = \begin{bmatrix} 1 & 0 & 0 \\ 0 & 1 & 0 \end{bmatrix} \begin{bmatrix} s & 0 & -1 \\ \frac{3}{2} & s+2 & \frac{1}{2} \\ 3 & 0 & s+4 \end{bmatrix}^{-1} \begin{bmatrix} 1 & 1 \\ -1 & -1 \\ -1 & -3 \end{bmatrix} + \begin{bmatrix} 1 & 0 \\ 1 & 1 \end{bmatrix}$$

$$= \begin{bmatrix} \frac{s+2}{s+1} & \frac{1}{s+3} \\ \frac{s}{s+1} & \frac{s+1}{s+2} \end{bmatrix}.$$

We can also write out the realization of controllability canonical form $\Sigma = (A_0, B_0, C_0)$:

$$A_0 = \begin{bmatrix} 0 & 0 & 0 & 0 & -6 & 0 \\ 0 & 0 & 0 & 0 & 0 & -6 \\ 1 & 0 & 0 & 0 & -11 & 0 \\ 0 & 1 & 0 & 0 & 0 & -11 \\ 0 & 0 & 1 & 0 & -6 & 0 \\ 0 & 0 & 0 & 1 & 0 & -6 \end{bmatrix}, \qquad B_0 = \begin{bmatrix} 6 & 2 \\ -6 & -3 \\ 5 & 3 \\ -5 & -4 \\ 1 & 1 \\ -1 & -1 \end{bmatrix},$$

$$C_0 = \begin{bmatrix} 0 & 0 & 0 & 0 & 1 & 0 \\ 0 & 0 & 0 & 0 & 0 & 1 \end{bmatrix}.$$

Then we decompose $\Sigma = (A_0, B_0, C_0)$ by controllability, then choose the transform matrix R_c according to equation (5.88):

$$R_c = \begin{bmatrix} 6 & 2 & -6 & 1 & 0 & 0 \\ -6 & -3 & 6 & 0 & 1 & 0 \\ 5 & 3 & -9 & 0 & 0 & 1 \\ -5 & -4 & 8 & 0 & 0 & 0 \\ 1 & 1 & -3 & 0 & 0 & 0 \\ -1 & -1 & 2 & 0 & 0 & 0 \end{bmatrix}.$$

Now we have:

$$R_c^{-1} = \begin{bmatrix} 0 & 0 & 0 & -1 & 0 & 4 \\ 0 & 0 & 0 & 1 & -2 & -7 \\ 0 & 0 & 0 & 0 & -1 & -1 \\ 1 & 0 & 0 & 4 & -2 & -16 \\ 0 & 1 & 0 & -3 & 0 & 9 \\ 0 & 0 & 1 & 2 & -3 & 8 \end{bmatrix}.$$

Thus:

$$\tilde{A} = R_c^{-1} A_0 R_c = \begin{bmatrix} \tilde{A}_{11} & \tilde{A}_{12} \\ 0 & \tilde{A}_{22} \end{bmatrix} = \begin{bmatrix} 1 & 0 & 0 & 0 & -1 & 0 \\ 0 & 0 & -6 & 0 & 1 & -2 \\ 0 & 1 & -5 & 0 & 0 & -1 \\ \hline 0 & 0 & 0 & 0 & 4 & -2 \\ 0 & 0 & 0 & 0 & -3 & 0 \\ 0 & 0 & 0 & 1 & 2 & -3 \end{bmatrix},$$

$$\tilde{B} = R_c^{-1} B_0 = \begin{bmatrix} \tilde{B}_1 \\ 0 \end{bmatrix} = \begin{bmatrix} 1 & 0 \\ 0 & 1 \\ 0 & 0 \\ 0 & 0 \\ 0 & 0 \\ 0 & 0 \end{bmatrix},$$

$$\tilde{C} = C_0 R_c = \begin{bmatrix} \tilde{C}_1 & 0 \end{bmatrix} = \begin{bmatrix} 1 & 1 & -3 & 0 & 0 & 0 \\ -1 & -1 & 2 & 0 & 0 & 0 \end{bmatrix}.$$

$\Sigma = (\tilde{A}_{11}, \tilde{B}_1, \tilde{C}_1)$ is a controllable and observable subsystem, so the minimum realization of $W(s)$ is:

$$\tilde{A}_m = \tilde{A}_{11} = \begin{bmatrix} 1 & 0 & 0 \\ 0 & 0 & -6 \\ 0 & 1 & -5 \end{bmatrix}, \qquad \tilde{B}_m = \tilde{B}_1 = \begin{bmatrix} 1 & 0 \\ 0 & 1 \\ 0 & 0 \end{bmatrix},$$

$$\tilde{C}_m = \tilde{C}_1 = \begin{bmatrix} 1 & 1 & -3 \\ -1 & -1 & 2 \end{bmatrix}, \qquad D = \begin{bmatrix} 1 & 0 \\ 1 & 1 \end{bmatrix}.$$

From the above calculation, we can see that if a transfer function is realizable, then it has an infinite number of realizations, although not necessarily of the same dimension. However, we can prove that if $\Sigma(A_m, B_m, C_m)$ and $\Sigma(\tilde{A}_m, \tilde{B}_m, \tilde{C}_m)$ are the minimum realizations of the same transfer matrix $W(s)$, then a state transformation $x = P\tilde{x}$ must exist, such that:

$$\tilde{A}_m = P^{-1} A_m P, \quad \tilde{B}_m = P^{-1} B_m, \quad \tilde{C}_m = C_m P.$$

We can see that the minimum realizations of the same transfer matrix are equivalent in algebra.

Example 5.21. Try to find the minimum realization of the transfer matrix using MAT-LAB:

$$W(s) = \begin{bmatrix} \frac{4s-10}{2s+1} & \frac{3}{s+2} \\ \frac{1}{(2s+1)(s+2)} & \frac{s+1}{(s+2)^2} \end{bmatrix}.$$

Solution. The controllability canonical form of the system is:

$$A = \begin{bmatrix} -4.5 & 0 & -6 & 0 & -2 & 0 \\ 0 & -4.5 & 0 & -6 & 0 & -2 \\ 1 & 0 & 0 & 0 & 0 & 0 \\ 0 & 1 & 0 & 0 & 0 & 0 \\ 0 & 0 & 1 & 0 & 0 & 0 \\ 0 & 0 & 0 & 1 & 0 & 0 \end{bmatrix}, \quad B = \begin{bmatrix} 1 & 0 \\ 0 & 1 \\ 0 & 0 \\ 0 & 0 \\ 0 & 0 \\ 0 & 0 \end{bmatrix},$$

$$C = \begin{bmatrix} -6 & 3 & -24 & 7.5 & -24 & 3 \\ 0 & 1 & 0.5 & 1.5 & 1 & 0.5 \end{bmatrix}, \quad D = \begin{bmatrix} 2 & 0 \\ 0 & 0 \end{bmatrix}.$$

This six-dimensional realization is clearly not minimal realizations. It can be reduced to minimal realizations by calling the MATLAB function `minreal`. We type:

```
a=[-4.5 0 -6 0 -2 0;0 -4.5 0 -6 0 -2;1 0 0 0 0 0;0 1 0 0 0 0;
    0 0 1 0 0 0;0 0 0 1 0 0];
b=[1 0;0 1;0 0;0 0;0 0;0 0];
c=[-6 3 -24 7.5 -24 3;0 1 0.5 1.5 1 0.5];
d=[2 0;0 0];;
[am,bm,cm,dm]=minreal(a,b,c,d)
```

Yield

```
am =
      -1.3387     0.2185    -1.6003
       2.5335    -1.1599     4.8338
      -0.0007    -0.0002    -2.0014
bm =
      -0.2666     0.2026
       0.2513    -0.6119
      -0.0001     0.3483
cm =
      32.7210    10.8346     8.6137
       0.8143    -0.8632     1.8281
dm =
       2     0
       5     0
```

Thus, the minimum realization is expressed as:

$$\dot{x}(t) = \begin{bmatrix} -1.3387 & 0.2185 & -1.6003 \\ 2.5335 & -1.1599 & 4.8338 \\ -0.0007 & -0.0002 & -2.0014 \end{bmatrix} x + \begin{bmatrix} -0.2666 & 0.2026 \\ 0.2513 & -0.6119 \\ -0.0001 & 0.3483 \end{bmatrix} u$$

$$y(t) = \begin{bmatrix} 32.7210 & 10.8346 & 8.6137 \\ -0.8143 & -0.8632 & 1.8281 \end{bmatrix} x + \begin{bmatrix} 2 & 0 \\ 0 & 0 \end{bmatrix} u .$$

5.7 Summary

The controllability and observability are both important properties of a system, and the definitions and criteria are described separately. The duality system is an important conception; duality systems have a series of interesting properties. The decomposition of a system is analyzed based on controllability and observability, and the minimum realization can be obtained accordingly.

Exercise

5.1. Check the controllability of the following systems:

(1) $\begin{bmatrix} \dot{x}_1 \\ \dot{x}_2 \end{bmatrix} = \begin{bmatrix} 1 & 1 \\ 1 & 0 \end{bmatrix} \begin{bmatrix} x_1 \\ x_2 \end{bmatrix} + \begin{bmatrix} 0 \\ 1 \end{bmatrix} u$

(2) $\begin{bmatrix} \dot{x}_1 \\ \dot{x}_2 \\ \dot{x}_3 \end{bmatrix} = \begin{bmatrix} 0 & 1 & 0 \\ 0 & 0 & 1 \\ -2 & -4 & -3 \end{bmatrix} \begin{bmatrix} x_1 \\ x_2 \\ x_3 \end{bmatrix} + \begin{bmatrix} 1 & 0 \\ 0 & 1 \\ -1 & 1 \end{bmatrix} \begin{bmatrix} u_1 \\ u_2 \end{bmatrix}$

(3) $\begin{bmatrix} \dot{x}_1 \\ \dot{x}_2 \\ \dot{x}_3 \end{bmatrix} = \begin{bmatrix} -3 & 1 & 0 \\ 0 & -3 & 0 \\ 0 & 0 & -1 \end{bmatrix} \begin{bmatrix} x_1 \\ x_2 \\ x_3 \end{bmatrix} + \begin{bmatrix} 1 & -1 \\ 0 & 0 \\ 2 & 0 \end{bmatrix} \begin{bmatrix} u_1 \\ u_2 \end{bmatrix}$

(4) $\begin{bmatrix} \dot{x}_1 \\ \dot{x}_2 \\ \dot{x}_3 \\ \dot{x}_4 \end{bmatrix} = \begin{bmatrix} \lambda_1 & 1 & 0 & 0 \\ 0 & \lambda_1 & 0 & 0 \\ 0 & 0 & \lambda_1 & 0 \\ 0 & 0 & 0 & \lambda_1 \end{bmatrix} \begin{bmatrix} x_1 \\ x_2 \\ x_3 \\ x_4 \end{bmatrix} + \begin{bmatrix} 0 \\ 1 \\ 1 \\ 1 \end{bmatrix} u$

(5) $\begin{bmatrix} \dot{x}_1 \\ \dot{x}_2 \\ \dot{x}_3 \end{bmatrix} = \begin{bmatrix} 0 & 4 & 3 \\ 0 & 20 & 16 \\ 0 & -25 & -20 \end{bmatrix} \begin{bmatrix} x_1 \\ x_2 \\ x_3 \end{bmatrix} + \begin{bmatrix} -1 \\ 3 \\ 0 \end{bmatrix} u$

5.2. Check the observability of the following systems:

(1) $\begin{bmatrix} \dot{x}_1 \\ \dot{x}_2 \end{bmatrix} = \begin{bmatrix} 1 & 1 \\ 1 & 0 \end{bmatrix} \begin{bmatrix} x_1 \\ x_2 \end{bmatrix}$, $\qquad y = \begin{bmatrix} 1 & 1 \end{bmatrix} \begin{bmatrix} x_1 \\ x_2 \end{bmatrix}$

(2) $\begin{bmatrix} \dot{x}_1 \\ \dot{x}_2 \\ \dot{x}_3 \end{bmatrix} = \begin{bmatrix} 0 & 1 & 0 \\ 0 & 0 & 1 \\ -2 & -4 & -3 \end{bmatrix} \begin{bmatrix} x_1 \\ x_2 \\ x_3 \end{bmatrix}$, $\quad \begin{bmatrix} y_1 \\ y_2 \end{bmatrix} = \begin{bmatrix} 0 & 1 & -1 \\ 1 & 2 & 1 \end{bmatrix} \begin{bmatrix} x_1 \\ x_2 \\ x_3 \end{bmatrix}$

(3) $\begin{bmatrix} \dot{x}_1 \\ \dot{x}_2 \\ \dot{x}_3 \end{bmatrix} = \begin{bmatrix} 0 & 4 & 3 \\ 0 & 20 & 16 \\ 0 & -25 & -20 \end{bmatrix} \begin{bmatrix} x_1 \\ x_2 \\ x_3 \end{bmatrix}$, $\quad y = \begin{bmatrix} -1 & 3 & 0 \end{bmatrix} \begin{bmatrix} x_1 \\ x_2 \\ x_3 \end{bmatrix}$

(4) $\begin{bmatrix} \dot{x}_1 \\ \dot{x}_2 \\ \dot{x}_3 \end{bmatrix} = \begin{bmatrix} 2 & 1 & 0 \\ 0 & 2 & 0 \\ 0 & 0 & -3 \end{bmatrix} \begin{bmatrix} x_1 \\ x_2 \\ x_3 \end{bmatrix}$, $\quad y = \begin{bmatrix} 0 & 1 & 1 \end{bmatrix} \begin{bmatrix} x_1 \\ x_2 \\ x_3 \end{bmatrix}$

(5) $\begin{bmatrix} \dot{x}_1 \\ \dot{x}_2 \\ \dot{x}_3 \end{bmatrix} = \begin{bmatrix} -4 & 0 & 0 \\ 0 & -4 & 0 \\ 0 & 0 & 1 \end{bmatrix} \begin{bmatrix} x_1 \\ x_2 \\ x_3 \end{bmatrix}$, $\quad y = \begin{bmatrix} 1 & 1 & 4 \end{bmatrix} \begin{bmatrix} x_1 \\ x_2 \\ x_3 \end{bmatrix}$

5.3. Is it possible to find a set of p and q, such that the state equation

$$\begin{bmatrix} \dot{x}_1 \\ \dot{x}_2 \end{bmatrix} = \begin{bmatrix} 1 & 12 \\ 1 & 0 \end{bmatrix} \begin{bmatrix} x_1 \\ x_2 \end{bmatrix} + \begin{bmatrix} p \\ -1 \end{bmatrix} u$$

$$y = \begin{bmatrix} q & 1 \end{bmatrix} \begin{bmatrix} x_1 \\ x_2 \end{bmatrix}$$

is not controllable/observable?

5.4. Try to prove that the system,

$$\begin{bmatrix} \dot{x}_1 \\ \dot{x}_2 \\ \dot{x}_3 \end{bmatrix} = \begin{bmatrix} 20 & -1 & 0 \\ 4 & 16 & 0 \\ 12 & 0 & 18 \end{bmatrix} \begin{bmatrix} x_1 \\ x_2 \\ x_3 \end{bmatrix} + \begin{bmatrix} a \\ b \\ c \end{bmatrix} u ,$$

is not controllable, no matter what a, b and c are.

5.5. Try to transform the following state space equation into controllability canonical form I.

$$\dot{x} = \begin{bmatrix} 1 & -2 \\ 3 & 4 \end{bmatrix} x + \begin{bmatrix} 1 \\ 1 \end{bmatrix} u$$

5.6. Try to transform the following state space equation into controllability canonical form I.

$$\dot{x} = \begin{bmatrix} -1 & -2 & -2 \\ 0 & -1 & 1 \\ 1 & 0 & 1 \end{bmatrix} x + \begin{bmatrix} 2 \\ 0 \\ 1 \end{bmatrix} u$$

5.7. Try to transform the following state space equation into observability canonical form II.

$$\dot{x} = \begin{bmatrix} -1 & -2 & -2 \\ 0 & -1 & 1 \\ 1 & 0 & 1 \end{bmatrix} x + \begin{bmatrix} 2 \\ 0 \\ 1 \end{bmatrix} u ,$$

$$y = \begin{bmatrix} 1 & 1 & 0 \end{bmatrix} x$$

5.8. Try to transform the following state space equation into observability canonical form II.

$$\dot{x} = \begin{bmatrix} 1 & -1 \\ 1 & 1 \end{bmatrix} x + \begin{bmatrix} 2 \\ 1 \end{bmatrix} u ,$$

$$y = \begin{bmatrix} -1 & 1 \end{bmatrix} x$$

5.9. Is the state equation

$$\dot{x} = \begin{bmatrix} -1 & 1 \\ 0 & 0 \end{bmatrix} x + \begin{bmatrix} 1 \\ 1 \end{bmatrix} u$$

controllable? If not, please give the controllability decomposition of the system.

5.10. Is the state equation

$$\dot{x}(t) = \begin{bmatrix} 1 & 2 & -1 \\ 0 & 1 & 0 \\ 1 & -4 & 3 \end{bmatrix} x(t) + \begin{bmatrix} 0 \\ 0 \\ 1 \end{bmatrix} u(t)$$

$$y(t) = \begin{bmatrix} 1 & -1 & 1 \end{bmatrix} x(t)$$

controllable? If not, please give the controllability decomposition of the system.

5.11. Is the state equation

$$\dot{x}(t) = \begin{bmatrix} 1 & 2 & -1 \\ 0 & 1 & 0 \\ 1 & -4 & 3 \end{bmatrix} x(t) + \begin{bmatrix} 0 \\ 0 \\ 1 \end{bmatrix} u(t)$$

$$y(t) = \begin{bmatrix} 1 & -1 & 1 \end{bmatrix} x(t)$$

observable? If not, please give the observability decomposition of the system.

5.12. Try to find the minimum realization of the transfer matrix:

$$W(s) = \begin{bmatrix} \frac{s+1}{s+2} \\ \frac{s+3}{(s+2)(s+4)} \end{bmatrix} .$$

5.13. The state equation of the inverted pendulum was developed in Example 1.10. Suppose, for a given pendulum, the equation becomes:

$$\dot{x} = \begin{bmatrix} 0 & 1 & 0 & 0 \\ 0 & 0 & -1 & 0 \\ 0 & 0 & 0 & 1 \\ 0 & 0 & 5 & 0 \end{bmatrix} x + \begin{bmatrix} 0 \\ 1 \\ 0 \\ -2 \end{bmatrix} u .$$

$$y = \begin{bmatrix} 1 & 0 & 0 & 0 \end{bmatrix} x$$

If $x_3 = \theta$ deviates from zero slightly, can we find a control u to push it back to zero? Why?

6 State Feedback and Observer

6.1 Introduction

Generally, the control theory can be divided into two parts; system analysis and system synthesis. In previous chapters, the solution of state space equations, the stability analysis and the controllability and observability of a control system were introduced. In this chapter, the system synthesis will be discussed. The controller is designed through feedback and the performance of a system is improved by pole assignment. The state estimator or state observer is designed to generate an estimation of the state.

6.2 Linear Feedback

Feedback is the most popular way to improve the performance of a system. Three kinds of linear feedback will be discussed in this chapter; the state feedback, the output feedback, and feedback from output y to $\dot{x}$. The algorithms are also described in detail.

6.2.1 State Feedback

Consider an n-dimensional LTI system:

$$\dot{x} = Ax + Bu$$
$$y = Cx + Du ,$$
(6.1)

where $x \in R^{n\times1}; u \in R^{p\times1}; y \in R^{m\times1}, A \in R^{n\times n}, B \in R^{n\times p}, C \in R^{m\times n}, D \in R^{m\times p}$.

Assume $D = 0$ to simplify the discussion. In state feedback, the input u is given by:

$$u = Kx + v = v + \begin{bmatrix} k_1 & k_2 & \cdots & k_n \end{bmatrix} x = v + \sum_{i=1}^{n} k_i x_i ,$$
(6.2)

where $v \in R^{p\times1}$ is the reference input and $K \in R^{p\times n}$ is the feedback gain matrix.

As shown in Figure 6.1, each feedback gain k_i is a real constant. This is called the constant gain negative state feedback, or state feedback for simplicity.

Substituting (6.2) into (6.1) yields the closed loop state space equation:

$$\dot{x} = (A + BK)x + Bv$$
$$y = Cx .$$
(6.3)

The closed loop transfer function is:

$$W_k(s) = C[sI - (A + BK)]^{-1}B .$$
(6.4)

https://doi.org/10.1515/9783110574951-006

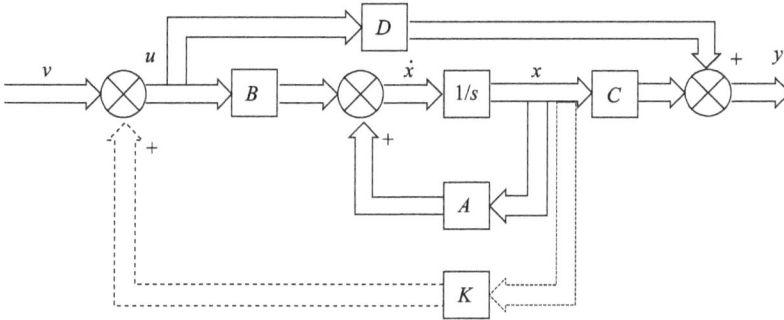

Fig. 6.1: State feedback.

6.2.2 Output Feedback

Output feedback is another linear feedback law using the output vector y, as shown in Figure 6.2.

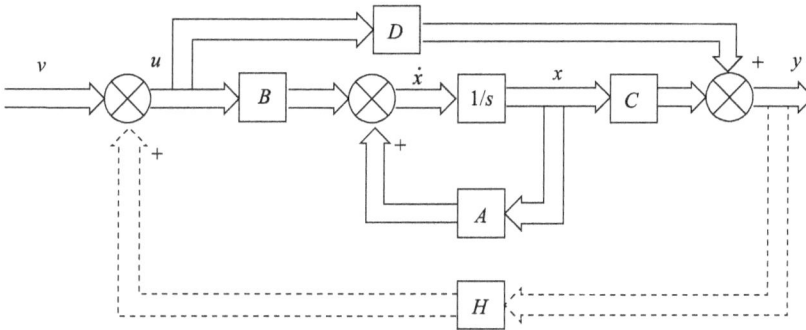

Fig. 6.2: Output feedback.

The control system is:

$$\dot{x} = Ax + Bu$$
$$y = Cx + Du \, .$$

$$(6.5)$$

The input u is given by:

$$u = Hy + v \, ,$$

$$(6.6)$$

where $H \in R^{p \times m}$ is the output feedback gain matrix.

If $D = 0$, the closed loop state space equation is:

$$\dot{x} = (A + BHC)x + Bv$$
$$y = Cx \, .$$

$$(6.7)$$

The closed loop transfer function is:

$$W_H(s) = C[sI - (A + BHC)]^{-1}B .$$ (6.8)

From equation (6.4), HC in output feedback is comparative to K in state feedback. As $m < n$, the optional degree of H is much smaller than that of K. Only when $C = I$ and $HC = K$, output feedback is absolutely equivalent to state feedback. Therefore, the effect of output feedback is obviously not as good as that of state feedback without compensator. However, output feedback shows its unique advantage in the facility in technique implementation.

6.2.3 Feedback From Output to $\dot{x}$

This linear feedback from system output y to state vector $\dot{x}$ is widely used in state estimator. This kind of feedback configuration is shown in Figure 6.3.

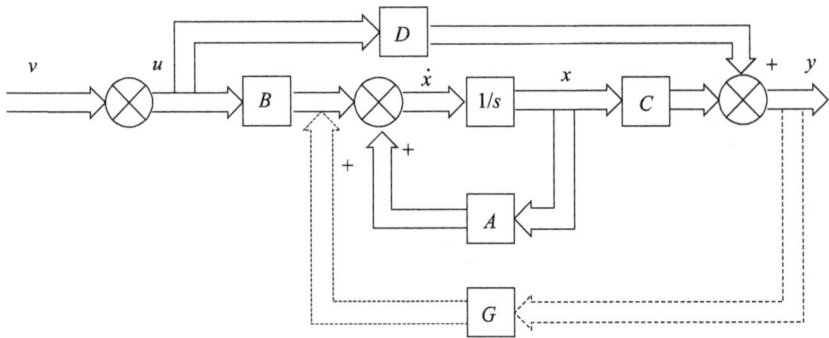

Fig. 6.3: Feedback from output y to $\dot{x}$.

The control system is:

$$\dot{x} = Ax + Bu$$
$$y = Cx + Du .$$ (6.9)

Considering the feedback gain matrix G, $G \in R^{n \times m}$ from the output y to the derivative of the state vector $\dot{x}$, the closed loop system is:

$$\dot{x} = Ax + Gy + Bu$$
$$y = Cx + Du .$$ (6.10)

Substituting y into $\dot{x}$ yields:

$$\dot{x} = (A + GC)x + (B + GD)u$$
$$y = Cx + Du .$$ (6.11)

If $D = 0$, then:

$$\dot{x} = (A + GC)x + Bu$$
$$y = Cx. \tag{6.12}$$

The closed loop transfer function is:

$$W_G(s) = C[sI - (A + GC)]^{-1}B. \tag{6.13}$$

As equation (6.13) shows, the change of matrix G will affect the eigenvalue of the closed loop system, and thus affect the characteristic of the system.

6.3 Pole Assignment

6.3.1 Sufficient and Necessary Condition for Arbitrary Pole Assignment

Theorem 6.1. *There exists a state feedback matrix K which can assign eigenvalues of the matrix A + BK of the closed loop system to arbitrary place of the state space, but only if the state vector of the open loop system is controllable. That is, if:*

$$\text{rank } Q_c = n,$$

where

$$Q_c = \begin{bmatrix} B & AB & A^2B & \cdots & A^{n-1}B \end{bmatrix}. \tag{6.14}$$

Proof (Only for Sufficiency). If the state vector of the open loop system is absolutely controllable, the following equation can be obtained with state feedback:

$$\det[\lambda I - (A + BK)] = f^*(\lambda), \tag{6.15}$$

where $f^*(\lambda)$ is the desired characteristic polynomial.

$$W_0(s) = C(sI - A)^{-1}B,$$

$$f^*(\lambda) = \prod_{i=1}^{n}(\lambda - \lambda_i^*) = \lambda^n + a_{n-1}^*\lambda^{n-1} + \cdots + a_1^*\lambda + a_0^*, \tag{6.16}$$

where λ_i^* ($i = 1, 2, \ldots, n$) are desired closed loop poles.

Step 1. If $\Sigma_0 = (A, B, C)$ is absolutely controllable, the following nonsingular transform exists:

$$x = T_{cl}\bar{x},$$

where T_{cl} is transfer matrix of controllable canonical form I.

Transform Σ_0 into the controllable canonical form I.

$$\dot{\bar{x}} = \bar{A}\bar{x} + \bar{B}u, \tag{6.17}$$

$$y = \bar{C}\bar{x},$$

where

$$\bar{A} = T_{cI}^{-1} A T_{cI} = \begin{bmatrix} 0 & 1 & 0 & \cdots & 0 \\ 0 & 0 & 1 & \cdots & 0 \\ \vdots & \vdots & \vdots & & \vdots \\ 0 & 0 & 0 & \cdots & 1 \\ -a_0 & -a_1 & -a_2 & \cdots & -a_{n-1} \end{bmatrix},$$

$$\bar{B} = T_{cI}^{-1} B = \begin{bmatrix} 0 \\ \vdots \\ 0 \\ 1 \end{bmatrix},$$

$$\bar{C} = C T_{cI} = \begin{bmatrix} b_0 & b_1 & \cdots & b_{n-1} \end{bmatrix}.$$

The transfer function of the controllable system Σ_0 is:

$$W_0(s) = C(sI - A)^{-1} B = \frac{b_{n-1}s^{n-1} + b_{n-2}s^{n-2} + \cdots + b_1 s + b_0}{s^n + a_{n-1}s^{n-1} + \cdots + a_1 s + a_0}. \tag{6.18}$$

Step 2. Consider the following state feedback gain matrix:

$$\bar{K} = \begin{bmatrix} \bar{k}_0 & \bar{k}_1 & \cdots & \bar{k}_{n-1} \end{bmatrix}. \tag{6.19}$$

Then we can obtain the closed loop state space description to $\bar{x}$:

$$\dot{\bar{x}} = (\bar{A} + \bar{B}\bar{K})\bar{x} + \bar{B}u, \tag{6.20}$$

$$y = \bar{C}\bar{x}.$$

Here,

$$\bar{A} + \bar{B}\bar{K} = \begin{bmatrix} 0 & 1 & 0 & \cdots & 0 \\ 0 & 0 & 1 & \cdots & 0 \\ \vdots & \vdots & \vdots & & \vdots \\ 0 & 0 & 0 & \cdots & 1 \\ -(a_0 - \bar{k}_0) & -(a_1 - \bar{k}_1) & \cdots & \cdots & -(a_{n-1} - \bar{k}_{n-1}) \end{bmatrix}.$$

Closed loop characteristic polynomial:

$$f(\lambda) = \left| \lambda I - (\bar{A} + \bar{B}\bar{K}) \right|$$

$$= \lambda^n + (a_{n-1} - \bar{k}_{n-1})\lambda^{n-1} + \cdots + (a_1 - \bar{k}_1)\lambda + (a_0 - \bar{k}_0). \tag{6.21}$$

Closed loop transfer function:

$$W_k(s) = \overline{C}\left[sI - (\overline{A} + \overline{B}\overline{K})\right]^{-1}\overline{B}$$

$$= \frac{b_{n-1}s^{n-1} + b_{n-2}s^{n-2} + \cdots + b_1 s + b_0}{s^n + (a_{n-1} - \overline{k}_{n-1})s^{n-1} + \cdots + (a_1 - \overline{k}_1)s + (a_0 - \overline{k}_0)} . \tag{6.22}$$

Step 3. To accord with the desired poles, the following equation has to be satisfied:

$$f(\lambda) = f^*(\lambda) .$$

The coefficients of the feedback matrix can be obtained by equaling the coefficients of λ with the same order in both sides of the above equation:

$$\overline{k}_i = a_i - a_i^* . \tag{6.23}$$

Thus,

$$\overline{K} = \begin{bmatrix} a_0 - a_0^* & a_1 - a_1^* & \cdots & a_{n-1} - a_{n-1}^* \end{bmatrix} .$$

Step 4. Transform $\overline{K}$ corresponding with $\overline{x}$ into K corresponding with x by using the following equation:

$$K = \overline{K}T_{cI}^{-1} . \tag{6.24}$$

This is due to $u = v + \overline{K}\overline{x} = v + KT_{cI}^{-1}x$. □

Theorem 6.2. *For the case of pole assignment via output feedback, wherein $u = Hy + v$, a theorem similar to Theorem 6.1 has not yet been proven. The determination of the output feedback matrix H is, in general, a very difficult task. A method for determining the matrix H, which is closely related to the method of determining the matrix K presented earlier, is based on the equation:*

$$K = HC . \tag{6.25}$$

This method starts with the determination of the matrix K and, in the sequel, the matrix H is determined by using equation (6.25). It is fairly easy to determine the matrix H from equation (6.25) since this equation is linear in H. Note that equation (6.25) is only a sufficient condition. That is, if equation (6.25) does not have a solution for H, it does not follow that pole assignment by output feedback is impossible.

Theorem 6.3. *Even for the absolutely controllable SISO system $\Sigma_o = (A, b, c)$, arbitrary pole assignment via output feedback cannot be guaranteed.*

Proof. The closed loop transfer function of the SISO feedback system, $\Sigma_h = [(A + bhc), b, c]$, is:

$$W_h(s) = c\left[sI - (A + bhc)\right]^{-1}b = \frac{W_o(s)}{1 + hW_o(s)} , \tag{6.26}$$

where $W_o(s) = c(sI - A)^{-1}b$ is the transfer function of the controllable system.

From the closed loop characteristic polynomial, we can obtain the closed loop root locus equation:

$$hW_0(s) = -1 . \tag{6.27}$$

When $W_0(s)$ is fixed beforehand, we can obtain a series of root locus with the reference variable h varying from 0 to ∞. Obviously, no matter what h is, the desired pole which is not contained in the root locus cannot be assigned. □

Output linear feedback has an important drawback: arbitrary pole assignment is not realizable.

To overcome the drawback, we always introduce additional regulatory network to affect the root locus by increasing open loop poles and zeros. Therefore, the regulated root locus consists of the desired pole.

Theorem 6.4. *For an absolutely controllable SISO system $\Sigma_0 = (A, b, c)$, the following is a sufficient and necessary condition of arbitrary pole assignment by the output feedback with dynamic compensator.*
(i) *Σ_0 is absolutely observable*
(ii) *the order of dynamic compensator is $n = 1$*

6.3.2 Methods to Assign the Poles of a System

1. Pole assignment via state feedback
Consider an LTI system

$$\dot{x}(t) = Ax(t) + Bu(t) \tag{6.28}$$

where we assume that all states are accessible and known. To this system, we apply a linear state feedback control law of the following form:

$$u(t) = -Kx(t) . \tag{6.29}$$

Then the closed loop system is given by the homogeneous equation:

$$\dot{x}(t) = (A - BK)x(t) . \tag{6.30}$$

It is remarked that the feedback law $u(t) = -Kx(t)$ is used rather than the feedback law $u(t) = Kx(t)$. This different chosen sign is utilized to facilitate the observer design problem.

Here, the design problem is to find the appropriate controller matrix K so as to improve the performance of the closed loop system (6.30). One method to improve the performance of (6.30) is pole assignment. The pole assignment method consists of finding a particular matrix K, such that the poles of the closed loop system (6.30) are set to desirable preassigned values. Using this method, the behavior of the open loop system may be improved significantly. For example, the method can stabilize an unstable system, increase or decrease the speed of response, widen or narrow the system's bandwidth, and increase or decrease the steady state error, etc. For these reasons, improving the system performance via the pole assignment method is widely used in practice.

The pole assignment, or eigenvalue assignment, problem can be defined as follows: suppose $\lambda_1, \lambda_2, \ldots, \lambda_n$ are the eigenvalues of the matrix A of the open loop system (6.30) and $\lambda_1^*, \lambda_2^*, \ldots, \lambda_n^*$ are the desired eigenvalues of matrix $A - BK$ of the closed loop system (6.30), where all complex eigenvalues appear in complex conjugate pairs. Denote $f(\lambda)$ and $f^*(\lambda)$ as the characteristic polynomial and the desired characteristic polynomial to find a matrix K so that equation (6.32) is satisfied.

$$f(\lambda) = \prod_{i=1}^{n}(\lambda - \lambda_i) = |\lambda I - A| = \lambda^n + a_{n-1}\lambda^{n-1} + \cdots + a_1\lambda + a_0 . \tag{6.31}$$

$$f^*(\lambda) = \prod_{i=1}^{n}(\lambda - \lambda_i^*) = |\lambda I - A + BK| = \lambda^n + a_{n-1}^*\lambda^{n-1} + \cdots + a_1^*\lambda + a_0^* . \tag{6.32}$$

The pole assignment problem has attracted considerable attention for many years. The first significant results were established by Wonham in the late 1960s and are given by Theorem 6.1 in Section 6.3.1.

According to Theorem 6.1, in cases that the open loop system (6.30) is not controllable, at least one eigenvalue of the matrix A remains invariant under the state feedback law (6.31). In such cases, in order to assign all eigenvalues, one must search for an appropriate dynamic controller wherein the feedback law (6.31) may involve, not only proportion, but also derivative, integral, and other terms. Dynamic controllers have the disadvantage in that they increase the order of the system.

Now, consider the case that the system (A, B) is controllable; a fact which guarantees that there exists a K which satisfies the pole assignment problem. Next, we will deal with the problem of determining a feedback matrix K. For simplicity, we will first study the case of single input systems, in which the matrix B reduces to a vector b and the matrix K reduces to a row vector k. Equation (6.32) then becomes:

$$f^*(\lambda) = \prod_{i=1}^{n}(\lambda - \lambda_i^*) = |\lambda I - A + bk| = \lambda^n + a_{n-1}^*\lambda^{n-1} + \cdots + a_1^*\lambda + a_0^* . \tag{6.33}$$

It has been remarked that the solution of equation (6.33) for k is unique.

Several methods have been proposed for determining k. We present three well known methods.

Method 1: The Base–Gura Formula
One of the most popular pole assignment methods gives the following simple solution:

$$k = -\overline{K}T_{\text{cI}}^{-1} . \tag{6.34}$$

Where $\bar{K}$ is defined in equation (6.23) and

$$T_{cI} = \begin{bmatrix} A^{n-1}b, & \ldots & Ab, & b \end{bmatrix} \begin{bmatrix} 1 & & & \\ a_{n-1} & \ddots & & \\ \vdots & & \ddots & \\ a_1 & \ldots & a_{n-1} & 1 \end{bmatrix}. \tag{6.35}$$

Method 2: The Phase Canonical Formula

Consider the special case that the system under control is described in its phase variable canonical form, i.e., A and b have the special forms A^* and b^*, where:

$$A^* = \begin{bmatrix} 0 & 1 & 0 & \ldots & 0 \\ 0 & 0 & 1 & \ldots & 0 \\ 0 & 0 & 0 & \ldots & 0 \\ \vdots & \vdots & \vdots & & \vdots \\ 0 & 0 & 0 & \ldots & 1 \\ -\hat{a}_0 & -\hat{a}_1 & -\hat{a}_2 & \ldots & -\hat{a}_{n-1} \end{bmatrix}, \quad b^* = \begin{bmatrix} 0 \\ 0 \\ 0 \\ \vdots \\ 0 \\ 1 \end{bmatrix}. \tag{6.36}$$

One of the most popular pole assignment methods gives the following simple solution:

$$k^T = [W^T Q_c^T]^{-1}(a^* - a). \tag{6.37}$$

Where Q_c is defined in equation (6.14) and

$$W = \begin{bmatrix} 1 & a_{n-1} & \ldots & a_1 \\ 0 & 1 & \ldots & a_2 \\ \vdots & \vdots & & \vdots \\ 0 & 0 & \ldots & 1 \end{bmatrix}, \quad a^* = \begin{bmatrix} a^*_{n-1} \\ a^*_{n-2} \\ \vdots \\ a^*_0 \end{bmatrix}, \quad a = \begin{bmatrix} a_{n-1} \\ a_{n-2} \\ \vdots \\ a_0 \end{bmatrix}. \tag{6.38}$$

Then, it can be easily shown that:

$$Q_c^* = \begin{bmatrix} b^* & A^*b^* & A^{*2}b^* & \ldots & A^{*n-1}b^* \end{bmatrix}.$$

The product $W^T Q_c^{*T}$ reduces to the simple form:

$$W^T Q_c^{*T} = \tilde{I} = \begin{bmatrix} 0 & 0 & \ldots & 0 & 1 \\ 0 & 0 & \ldots & 1 & 0 \\ \vdots & \vdots & & \vdots & \vdots \\ 1 & 0 & \ldots & 0 & 0 \end{bmatrix}. \tag{6.39}$$

In this case, the vector k^{*T} in expression (6.37) reduces to $k^{*T} = \tilde{I}(a^* - a)$, i.e., it reduces to the following form:

$$k^{*T} = \tilde{I}(a^* - a) = \begin{bmatrix} a^*_0 - a_0 \\ a^*_1 - a_1 \\ \vdots \\ a^*_{n-1} - a_{n-1} \end{bmatrix}. \tag{6.40}$$

It is evident that expression (6.40) is extremely simple to apply, provided that the matrix A and the vector b of the system under control are in the phase variable canonical form (6.38).

Method 3: The Ackermann's Formula
Another approach for computing k has been proposed by Ackermann, which leads to the following expression:

$$k = e^T Q_c^{-1} f^*(A) . \tag{6.41}$$

The matrix Q_c is given in equation (6.14), wherein the variable s is replaced by the matrix A, i.e.,

$$f^*(A) = A^n + a_{n-1}^* A^{n-1} + \cdots + a_1^* A + a_0^* I . \tag{6.42}$$

In general cases of multi-input systems, the determination of the matrix K is somewhat complicated. A simple approach to the problem is to assume that K has the following form:

$$K = q p^T , \tag{6.43}$$

where q and p are n-dimensional vectors. Then, the matrix $A - BK$ becomes:

$$A - BK = A - B q p^T = A - \beta p^T , \quad \text{where} \quad \beta = Bq . \tag{6.44}$$

Therefore, assuming that K has the form of equation (6.43), the multi-input system is reduced to a single input system, which has been studied previously. In other words, the solution for the vector p is equation (6.37) or equation (6.41), and differs only in that the matrix Q_c is now the matrix $\widetilde{Q}_c$, which takes the following form:

$$\widetilde{Q}_c = \begin{bmatrix} \beta & A\beta & A^2\beta & \cdots & A^{n-1}\beta \end{bmatrix} , \quad K = HC . \tag{6.45}$$

The vector $\beta = Bq$ involves arbitrary parameters, which are the elements of the arbitrary vector q. These arbitrary parameters can have any value, provided that rank $\widetilde{Q}_c = n$.

2. Pole assignment via output feedback
Consider a LTI system:

$$\begin{aligned} \dot{x}(t) &= Ax(t) + Bu(t) \\ y(t) &= Cx(t) , \end{aligned} \tag{6.46}$$

where we assume that all states are accessible and known. We apply the following linear state feedback control law to the above system:

$$u(t) = -Hy(t) + v . \tag{6.47}$$

Then the closed loop system is given by the homogeneous equation:

$$\dot{x}(t) = (A - BHC)x(t) . \tag{6.48}$$

The pole assignment, or eigenvalue assignment, problem can be defined as follows: denote $\lambda_1, \lambda_2, \ldots, \lambda_n$ as the eigenvalues of the matrix A of the open loop system (6.46) and $\lambda_1^*, \lambda_2^*, \ldots, \lambda_n^*$ as the desired eigenvalues of matrix $A - BHC$ of the closed loop system (5.3)–(5.35), where all complex eigenvalues appear in complex conjugate pairs.

In the case of pole assignment via output feedback, wherein $u = -Hy + v$, Theorem 6.2 has been proven. According to Theorem 6.2, we can obtain the matrix H by $K = HC$.

This method starts with the determination of the matrix K, and in the following content, the matrix H is determined by using equation (6.25). It is fairly easy to determine the matrix H from equation (6.25) since this equation is linear in H. Note that equation (6.25) is only a sufficient condition, i.e., if equation (6.25) does not have a solution for H, it does not follow that pole assignment by output feedback is impossible.

6.3.3 Examples

Example 6.1. Consider a system in the form (6.28), where:

$$A = \begin{bmatrix} 0 & 1 \\ -1 & 0 \end{bmatrix} \quad \text{and} \quad b = \begin{bmatrix} 0 \\ 1 \end{bmatrix}.$$

Find a vector k to make the closed loop system eigenvalues become $\lambda_1^* = -1$ and $\lambda_2^* = -1.5$.

Solution. We have:

$$f(\lambda) = |\lambda I - A| = \lambda^2 + 1 \quad \text{and} \quad f^*(\lambda) = (\lambda - \lambda_1^*)(\lambda - \lambda_2^*) = \lambda^2 + 2.5\lambda + 1.5.$$

Method 1
Here, we use equation (6.35) and (6.23):

$$T_{cl} = \begin{bmatrix} Ab & b \end{bmatrix} \begin{bmatrix} 1 & 0 \\ 0 & 1 \end{bmatrix} = \begin{bmatrix} 1 & 0 \\ 0 & 1 \end{bmatrix}\begin{bmatrix} 1 & 0 \\ 0 & 1 \end{bmatrix} = \begin{bmatrix} 1 & 0 \\ 0 & 1 \end{bmatrix},$$

And:
$$\overline{K} = \begin{bmatrix} a_0 - a_0^* & a_1 - a_1^* \end{bmatrix} = \begin{bmatrix} 1 - 1.5 & 0 - 2.5 \end{bmatrix} = \begin{bmatrix} -0.5 & -2.5 \end{bmatrix}.$$

Therefore:
$$T_{cl}^{-1} = \begin{bmatrix} 1 & 0 \\ 0 & 1 \end{bmatrix}^{-1} = \begin{bmatrix} 1 & 0 \\ 0 & 1 \end{bmatrix}.$$

Hence:
$$k = -\overline{K}T_{cl}^{-1} = -\begin{bmatrix} -0.5 & -2.5 \end{bmatrix}\begin{bmatrix} 1 & 0 \\ 0 & 1 \end{bmatrix} = \begin{bmatrix} 0.5 & 2.5 \end{bmatrix}.$$

Method 2

Since the system is in phase variable canonical form, the vector k can readily be determined by equation (6.40), as follows:

$$k^T = k^{*T} = \begin{bmatrix} a_0^* - a_0 \\ a_1^* - a_1 \end{bmatrix} = \begin{bmatrix} 1.5 - 1 \\ 2.5 - 0 \end{bmatrix} = \begin{bmatrix} 0.5 \\ 2.5 \end{bmatrix}.$$

Method 3

Here, we apply equation (6.41). We have:

$$f^*(A) = A^2 + a_1^* A + a_0^* I = A^2 + 2.5A + 1.5I$$

$$= \begin{bmatrix} 0 & 1 \\ -1 & 0 \end{bmatrix}^2 + 2.5 \begin{bmatrix} 0 & 1 \\ -1 & 0 \end{bmatrix} + 1.5 \begin{bmatrix} 1 & 0 \\ 0 & 1 \end{bmatrix}$$

$$= \begin{bmatrix} -1 & 0 \\ 0 & -1 \end{bmatrix} + \begin{bmatrix} 0 & 2.5 \\ -2.5 & 0 \end{bmatrix} + \begin{bmatrix} 1.5 & 0 \\ 0 & 1.5 \end{bmatrix} = \begin{bmatrix} 0.5 & 2.5 \\ -2.5 & 0.5 \end{bmatrix}$$

$$Q_c^{-1} = \begin{bmatrix} b & Ab \end{bmatrix}^{-1} = \begin{bmatrix} 0 & 1 \\ 1 & 0 \end{bmatrix}^{-1} = \begin{bmatrix} 0 & 1 \\ 1 & 0 \end{bmatrix}.$$

Therefore:

$$k = e^T S^{-1} f^*(A) = \begin{bmatrix} 0 & 1 \end{bmatrix} \begin{bmatrix} 0 & 1 \\ 1 & 0 \end{bmatrix} \begin{bmatrix} 0.5 & 2.5 \\ -2.5 & 0.5 \end{bmatrix} = \begin{bmatrix} 0.5 & 2.5 \end{bmatrix}.$$

Clearly, the resulting three controller vectors derived by the three methods are identical. This is due to the fact that, for a single input system, k is unique.

Example 6.2. Consider a system in the form (6.28), where:

$$A = \begin{bmatrix} 0 & 1 & 0 \\ 0 & 0 & 1 \\ 1 & 0 & 0 \end{bmatrix} \quad \text{and} \quad b = \begin{bmatrix} 0 \\ 0 \\ 1 \end{bmatrix}.$$

Find a vector k to make the eigenvalues of the closed loop system into $\lambda_1^* = -1$, $\lambda_2^* = -2$, and $\lambda_3^* = -2$.

Solution. We have:

$$f(\lambda) = |\lambda I - A| = \lambda^3 - 1,$$

And:

$$f^*(\lambda) = (\lambda - \lambda_1^*)(\lambda - \lambda_2^*)(\lambda - \lambda_3^*) = \lambda^3 + 5\lambda^2 + 8\lambda + 4.$$

Method 1

Here we use equations (6.37) and (6.23):

$$T_{cI} = \begin{bmatrix} A^2 b & Ab & b \end{bmatrix} \begin{bmatrix} 1 & 0 & 0 \\ 0 & 1 & 0 \\ 0 & 0 & 1 \end{bmatrix} = \begin{bmatrix} 1 & 0 & 0 \\ 0 & 1 & 0 \\ 0 & 0 & 1 \end{bmatrix} \begin{bmatrix} 1 & 0 & 0 \\ 0 & 1 & 0 \\ 0 & 0 & 1 \end{bmatrix} = \begin{bmatrix} 1 & 0 & 0 \\ 0 & 1 & 0 \\ 0 & 0 & 1 \end{bmatrix},$$

And:

$$\overline{K} = \begin{bmatrix} a_0 - a_0^* & a_1 - a_1^* & a_2 - a_2^* \end{bmatrix} = \begin{bmatrix} -1 - 4 & 0 - 8 & 0 - 5 \end{bmatrix} = \begin{bmatrix} -5 & -8 & -5 \end{bmatrix}.$$

Therefore:

$$T_{cI}^{-1} = \begin{bmatrix} 1 & 0 & 0 \\ 0 & 1 & 0 \\ 0 & 0 & 1 \end{bmatrix}^{-1} = \begin{bmatrix} 1 & 0 & 0 \\ 0 & 1 & 0 \\ 0 & 0 & 1 \end{bmatrix}.$$

Hence:

$$k = -\overline{K}T_{cI}^{-1} = -\begin{bmatrix} -5 & -8 & -5 \end{bmatrix}\begin{bmatrix} 1 & 0 & 0 \\ 0 & 1 & 0 \\ 0 & 0 & 1 \end{bmatrix} = \begin{bmatrix} 5 & 8 & 5 \end{bmatrix}.$$

Method 2

Since the system is in phase variable canonical form, the vector k can readily be determined by equation (6.40), as follows:

$$k^{\mathrm{T}} = k^{*T} = \begin{bmatrix} a_0^* - a_0 \\ a_1^* - a_1 \\ a_2^* - a_2 \end{bmatrix} = \begin{bmatrix} 4 - (-1) \\ 8 - 0 \\ 5 - 0 \end{bmatrix} = \begin{bmatrix} 5 \\ 8 \\ 5 \end{bmatrix}.$$

Method 3

Here, we apply equation (6.41). We have:

$$f^*(A) = A^3 + a_2^* A^2 + a_1^* A + a_0^* I = A^3 + 5A^2 + 8A + 4I$$

$$= \begin{bmatrix} 0 & 1 & 0 \\ 0 & 0 & 1 \\ 1 & 0 & 0 \end{bmatrix}^3 + 5\begin{bmatrix} 0 & 1 & 0 \\ 0 & 0 & 1 \\ 1 & 0 & 0 \end{bmatrix}^2 + 8\begin{bmatrix} 0 & 1 & 0 \\ 0 & 0 & 1 \\ 1 & 0 & 0 \end{bmatrix} + 4\begin{bmatrix} 1 & 0 & 0 \\ 0 & 1 & 0 \\ 0 & 0 & 1 \end{bmatrix}$$

$$= \begin{bmatrix} 1 & 0 & 0 \\ 0 & 1 & 0 \\ 0 & 0 & 1 \end{bmatrix} + \begin{bmatrix} 0 & 0 & 5 \\ 5 & 0 & 0 \\ 0 & 5 & 0 \end{bmatrix} + \begin{bmatrix} 0 & 8 & 0 \\ 0 & 0 & 8 \\ 8 & 0 & 0 \end{bmatrix} + \begin{bmatrix} 4 & 0 & 0 \\ 0 & 4 & 0 \\ 0 & 0 & 4 \end{bmatrix}$$

$$= \begin{bmatrix} 5 & 8 & 5 \\ 5 & 5 & 8 \\ 8 & 5 & 5 \end{bmatrix}.$$

Therefore:

$$k = e^{\mathrm{T}}S^{-1}f^*(A) = \begin{bmatrix} 0 & 0 & 1 \end{bmatrix}\begin{bmatrix} 0 & 0 & 1 \\ 0 & 1 & 0 \\ 1 & 0 & 0 \end{bmatrix}\begin{bmatrix} 5 & 8 & 5 \\ 5 & 5 & 8 \\ 8 & 5 & 5 \end{bmatrix} = \begin{bmatrix} 5 & 8 & 5 \end{bmatrix}.$$

Example 6.3. Consider the following system with transfer function as:

$$W(s) = \frac{10}{s(s + 1)(s + 2)}.$$

Try to find a state feedback controller to make the closed loop poles become -2 and $-1 \pm j1$.

Solution. Since the system is controllable and observable, the poles can be assigned arbitrarily. Choose the following controllable canonical form:

$$\dot{x} = \begin{bmatrix} 0 & 1 & 0 \\ 0 & 0 & 1 \\ 0 & -2 & -3 \end{bmatrix} x + \begin{bmatrix} 0 \\ 0 \\ 1 \end{bmatrix} u,$$

$$y = \begin{bmatrix} 10 & 0 & 0 \end{bmatrix} x.$$

With state feedback, the closed loop characteristic polynomial is:

$$f(\lambda) = \det[\lambda I - (A + bK)] = \lambda^3 + (3 - k_2)\lambda^2 + (2 - k_1)\lambda - k_0.$$

The desired closed loop characteristic polynomial is:

$$f^*(\lambda) = (\lambda + 2)(\lambda + 1 - j)(\lambda + 1 + j) = \lambda^3 + 4\lambda^2 + 6\lambda + 4.$$

Compare relative parameters in the above two functions, and we have:

$$k_0 = -4, \quad k_1 = -4, \quad k_2 = -1.$$

Thus:

$$K = \begin{bmatrix} -4 & -4 & -1 \end{bmatrix}.$$

The closed loop transfer function is:

$$G(s) = \frac{10}{s^3 + 4s^2 + 6s + 4}.$$

The block diagram of the closed loop system is shown in Figure 6.4.

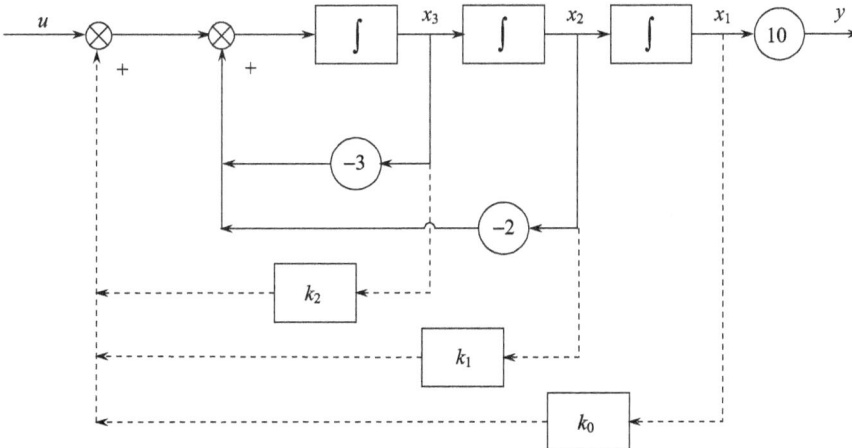

Fig. 6.4: Block diagram of the closed loop system.

Example 6.4. The state space model of a system is:

$$\dot{x} = \begin{bmatrix} 0 & 1 & 0 \\ 0 & -1 & 1 \\ 0 & 0 & -2 \end{bmatrix} x + \begin{bmatrix} 0 \\ 0 \\ 1 \end{bmatrix} u$$

$$y = \begin{bmatrix} 1 & 0 & 0 \end{bmatrix} x .$$

Try to find a state feedback controller to make the closed loop poles become -2 and $-1 \pm j1$.

Solution. To determine the controllability of the system:

$$M = \begin{bmatrix} b & Ab & A^2b \end{bmatrix} = \begin{bmatrix} 0 & 0 & 1 \\ 0 & 1 & -3 \\ 1 & -2 & -4 \end{bmatrix} .$$

$|M| \neq 0$, so the system is controllable, and the closed loop poles of the system can be arbitrary assigned.

Transform the above state space model into the controllable canonical form. The characteristic function is:

$$|sI - A| = s^3 + 3s^2 + 2s .$$

So we choose:

$$I = \begin{bmatrix} 2 & 3 & 1 \\ 3 & 1 & 0 \\ 1 & 0 & 0 \end{bmatrix} .$$

Then:

$$T_{c1} = MI = \begin{bmatrix} 1 & 0 & 0 \\ 0 & 1 & 0 \\ 0 & 1 & 1 \end{bmatrix}, \quad T_{c1}^{-1} = \begin{bmatrix} 1 & 0 & 0 \\ 0 & 1 & 0 \\ 0 & -1 & 1 \end{bmatrix} .$$

Suppose $\hat{k} = \begin{bmatrix} \hat{k}_0 & \hat{k}_1 & \hat{k}_2 \end{bmatrix}$ and that the closed loop characteristic polynomials can be expressed as:

$$f(\lambda) = |\lambda I - (\hat{A} + \hat{b}\hat{k})| = |\lambda I - (T^{-1}AT + T^{-1}b\hat{k})| .$$

They can also be expressed as:

$$f^*(\lambda) = (\lambda + 2)(\lambda + 1 - j)(\lambda + 1 + j) = \lambda^3 + 4\lambda^2 + 6\lambda + 4 .$$

To achieve the desired closed loop poles, we have $f^*(\lambda) = f(\lambda)$.
   They can also be expressed as:

$$k = \hat{k}T_{c1}^{-1}, \quad \text{so} \quad k = \begin{bmatrix} -4 & -4 & 1 \end{bmatrix} \begin{bmatrix} 1 & 0 & 0 \\ 0 & 1 & 0 \\ 0 & -1 & 1 \end{bmatrix} = \begin{bmatrix} -4 & -3 & -1 \end{bmatrix} .$$

The block diagram of the closed loop system is shown in Figure 6.5.

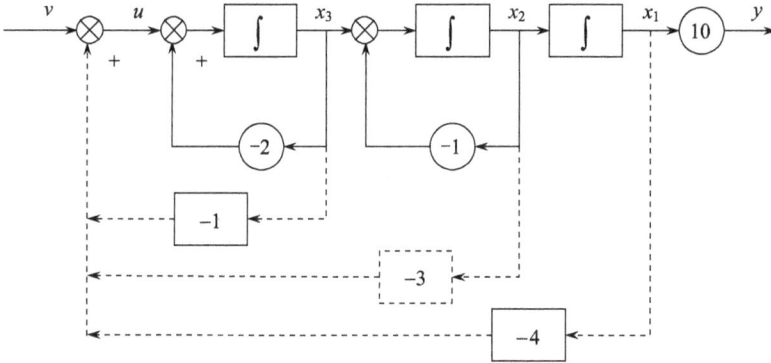

Fig. 6.5: Block diagram of the closed loop system.

Example 6.5. Consider a plant described by:

$$\dot{x} = \begin{bmatrix} 0 & 1 & 0 \\ 0 & 0 & 1 \\ 0 & -2 & -3 \end{bmatrix} x + \begin{bmatrix} 0 \\ 0 \\ 1 \end{bmatrix} u .$$

Let us introduce state feedback $u = r - [k_1 \ k_2 \ k_3]x$ to place the three eigenvalues at $-2, -1 \pm j$. Figure out how to solve it using MATLAB.

The MATLAB function place computes state feedback gains for eigenvalue placement or assignment. For example, we can type:

```
a=[0 1 0;0 0 1;0 -2 -3];b=[0;0;1];
p=[-2,-1+j,-1-j];
k=place(a,b,p)
```

yield

```
k =
    4.0000    4.0000    1.0000
```

This is the matrix $[k_1 \ k_2 \ k_3] = [4 \ 4 \ 1]$.

6.4 State Estimator

6.4.1 Introduction

In the preceding sections, we introduce state feedback under the implicit assumption that all state variables are available for feedback. This assumption may not hold in practice, either because the state variables are not accessible for direct connection or because sensors or transducers are not available. In this case, in order to apply state feedback, we must design a device, called *a state estimator or state observer*, so that the output of the device will generate an estimation of the state.

6.4.2 State Estimator

1. Full Dimensional State Estimator
Consider the LTI system $\Sigma_0 = (A, B, C)$:

$$\dot{x} = Ax + Bu$$
$$y = Cx ,$$

(6.49)

where A, B, and C are given and the input $u(t)$ and the output $y(t)$ are available to us. The problem is to estimate x from u and y with the knowledge of A, B, and C. If we know A and B, we can duplicate the original system as:

$$\dot{\hat{x}} = A\hat{x} + Bu ,$$

(6.50)

which is shown in Figure 6.6. The duplication will be called an open loop estimator. Now, if (6.49) and (6.50) have the same initial state, then for any input, we have $\hat{x}(t) = x(t)$ for all $t \geq 0$. Therefore, the remaining question is how to find the initial state of (6.49) and then set the initial state of (6.50) to that state. If (6.49) is observable, its initial state $x(0)$ can be computed from u and y over any time interval, say, $[0, t_1]$. We can then compute the state at t_2 and set $\hat{x}(t_2) = x(t_2)$. Then we have $\hat{x}(t) = x(t)$ for all $t \geq t_2$. Thus, if (6.49) is observable, an open loop estimator can be used to generate the state vector.

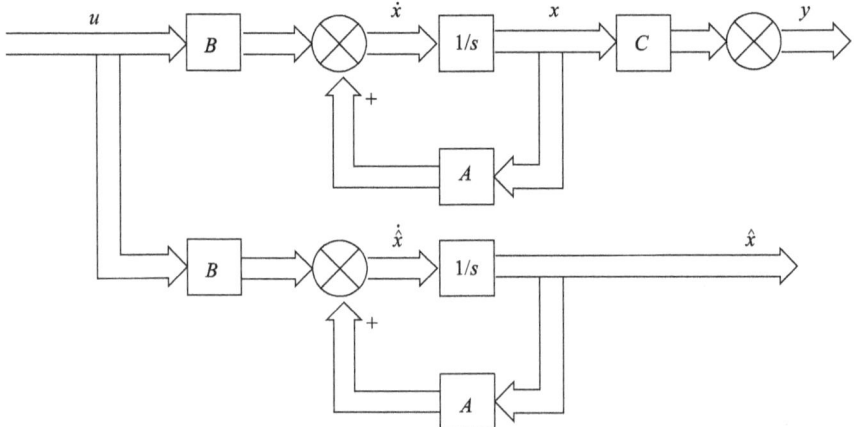

Fig. 6.6: Block diagram of open loop state estimator.

There are, however, two disadvantages, in using an open loop estimator. First, the initial state must be computed and set each time we use the estimator, which is very inconvenient. Secondly, and more seriously, if the matrix A has eigenvalues with positive real part, then even for a very small difference between $x(t_0)$ and $\hat{x}(t_0)$ for some t_0, which may be caused by disturbance or imperfect estimation of the initial state, the difference between $x(t)$ and $\hat{x}(t)$ will grow with time. Therefore, the open loop estimator is, in general, not satisfactory.

We see from Figure 6.6 that, even though the input and output of (6.49) are available, we use only the input to drive the estimator. Now we shall modify the estimator in Figure 6.6 to the one in Figure 6.7, in which the output $y(t) = Cx(t)$ of (6.49) is compared with $C\hat{x}(t)$. Their difference, passing through an $n \times 1$ constant gain vector G, is used as a correcting term. If the difference is zero, no correction is needed. If the difference is nonzero and if the gain G is properly designed, the difference will drive the estimated state to the actual state. Such an estimator is called a closed loop or an asymptotic estimator or, simply, an estimator.

The open loop estimator is now modified as:

$$\dot{\hat{x}} = A\hat{x} + Bu + G(y - \hat{y}) = A\hat{x} + Bu + Gy - GC\hat{x}, \qquad (6.51)$$

which is shown in Figure 6.7. Now, (6.51) can be written as:

$$\dot{\hat{x}} = (A - GC)\hat{x} + Bu + Gy, \qquad (6.52)$$

and is shown in Figure 6.8. It has two inputs, u and y, and its output yields an estimated state $\hat{x}$.

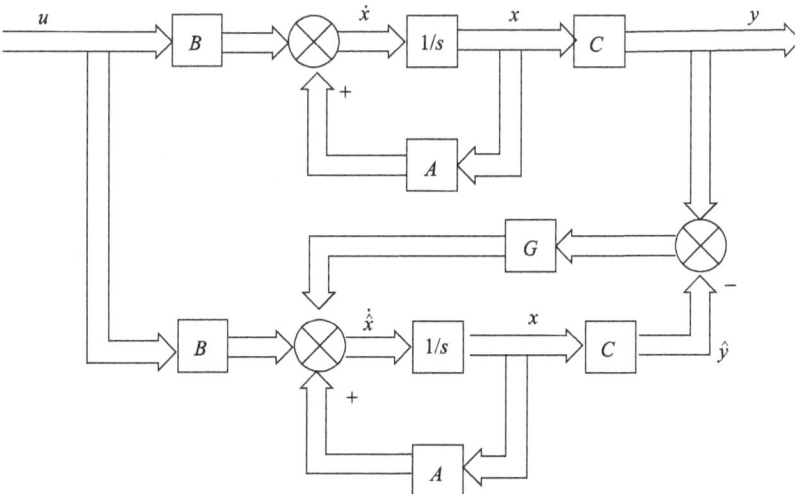

Fig. 6.7: Block diagram of closed loop state estimator I.

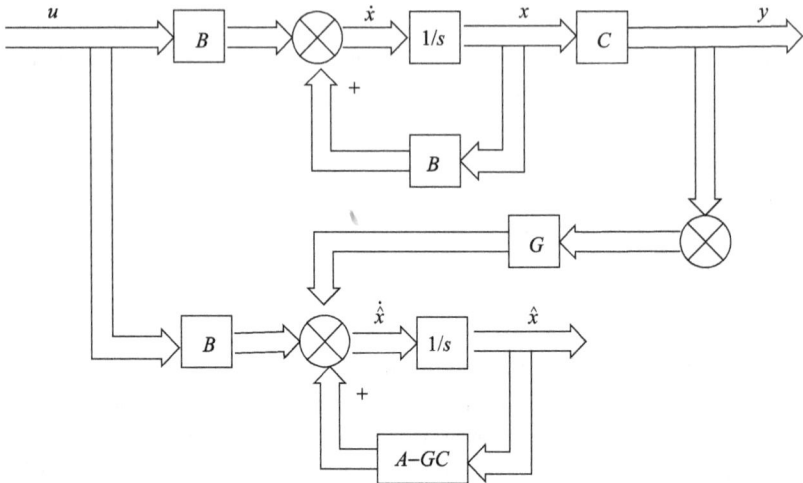

Fig. 6.8: Block diagram of closed loop state estimator Π.

Let us define:

$$e = x - \hat{x}.$$

It is the error between the actual state and estimated state. Differentiating e and then substituting (6.49) and (6.51) into it, we obtain:

$$\dot{e} = \dot{x} - \dot{\hat{x}} = Ax + Bu - (A - GC)\hat{x} - G(Cx) - Bu$$
$$= Ax - (A - GC)\hat{x} - GCx = (A - GC)(x - \hat{x}),$$

or

$$\dot{e} = (A - GC)e.\qquad\qquad (6.53)$$

This equation governs the estimation error. If all eigenvalues of $(A - GC)$ can be assigned arbitrarily, then we can control the rate for e to approach zero or, equivalently, for the estimated state to approach the actual state. For example, if all eigenvalues of $A - GC$ have negative real parts smaller than $-\sigma$, then the entire of e will approach zero at rates faster than $e^{-\sigma t}$. Therefore, even if there is a large error between $\hat{x}(t_0)$ and $x(t_0)$ at the initial time t_0, the estimated state will approach the actual state rapidly. Thus, there is no need to compute the initial state of the original state equation. In conclusion, if all eigenvalues of $(A - GC)$ are properly assigned, a closed loop estimator is much more desirable than an open loop estimator.

Theorem 6.5. *Consider the pair (A, C). All eigenvalues of $(A - GC)$ can be assigned arbitrarily by selecting a real constant vector G if, and only if, (A, C) is observable.*

Example 6.6. Consider the state equation:

$$\dot{x} = \begin{bmatrix} 1 & 0 \\ 0 & 0 \end{bmatrix} x + \begin{bmatrix} 1 \\ 1 \end{bmatrix} u \,,$$

$$y = \begin{bmatrix} 2 & -1 \end{bmatrix} x \,.$$

Try to find the state estimator, so that the desired closed loop eigenvalues can be $-10, -10$.

Solution. Step 1: Examine the system's observability. We have:

$$N = \begin{bmatrix} C \\ CA \end{bmatrix} = \begin{bmatrix} 2 & -1 \\ 2 & 0 \end{bmatrix} \,.$$

Since rank $N = 2$, there exists a full dimensional state estimator.

Step 2: Transfer the system to observability criterion form Π.
The characteristic polynomial of the system is:

$$\det [\lambda I - A] = \det \begin{bmatrix} \lambda - 1 & 0 \\ 0 & \lambda \end{bmatrix} = \lambda^2 - \lambda \,.$$

We have:

$$a_1 = -1 \,, \quad a_0 = 0 \,, \quad L = \begin{bmatrix} a_1 & 1 \\ 1 & 0 \end{bmatrix} = \begin{bmatrix} -1 & 1 \\ 1 & 0 \end{bmatrix} \,,$$

and:

$$T^{-1} = LN = \begin{bmatrix} -1 & 1 \\ 1 & 0 \end{bmatrix} \begin{bmatrix} 2 & -1 \\ 2 & 0 \end{bmatrix} = \begin{bmatrix} 0 & 1 \\ 2 & -1 \end{bmatrix} \,, \quad T = \begin{bmatrix} \frac{1}{2} & \frac{1}{2} \\ 1 & 0 \end{bmatrix} \,.$$

Thus:

$$\dot{\overline{x}} = T^{-1}AT\overline{x} + T^{-1}bu = \begin{bmatrix} 0 & 0 \\ 1 & 1 \end{bmatrix} \overline{x} + \begin{bmatrix} 1 \\ 1 \end{bmatrix} u \,,$$

$$y = CT\overline{x} \,.$$

Step 3: Introducing feedback matrix $\overline{G} = [\overline{g}_1 \ \overline{g}_2]^T$. The characteristic polynomial of the estimator yields:

$$f(\lambda) = |\lambda I - (\overline{A} - \overline{G}\overline{C})| = \begin{vmatrix} \lambda & \overline{g}_1 \\ -1 & \lambda - (1 - \overline{g}_2) \end{vmatrix} = \lambda^2 - (1 - \overline{g}_2)\lambda + \overline{g}_1 \,.$$

Step 4. The desired characteristic polynomial is:

$$f^*(\lambda) = (\lambda + 10)^2 = \lambda^2 + 20\lambda + 100 \,.$$

Step 5. Comparing the corresponding coefficient of $f(\lambda)$ and $f^*(\lambda)$, we have:

$$\overline{g}_1 = 100 \,, \quad \overline{g}_2 = 21 \,, \quad \text{and} \quad \overline{G} = \begin{bmatrix} 100 \\ 21 \end{bmatrix} \,.$$

Step 6. Transforming to the state of x yields:

$$G = T\overline{G} = \begin{bmatrix} \frac{1}{2} & \frac{1}{2} \\ 1 & 0 \end{bmatrix} \begin{bmatrix} 100 \\ 21 \end{bmatrix} = \begin{bmatrix} 60.5 \\ 100 \end{bmatrix}.$$

Step 7. The proposed estimator is:

$$\dot{\hat{x}} = (A - Gc)\hat{x} + bu + Gy$$

$$= \begin{bmatrix} -120 & 60.5 \\ -200 & 100 \end{bmatrix} \hat{x} + \begin{bmatrix} 1 \\ 1 \end{bmatrix} u + \begin{bmatrix} 60.5 \\ 100 \end{bmatrix} y,$$

Or:

$$\dot{\hat{x}} = A\hat{x} + bu + G(y - \hat{y}) = \begin{bmatrix} 1 & 0 \\ 0 & 0 \end{bmatrix} \hat{x} + \begin{bmatrix} 1 \\ 1 \end{bmatrix} u + \begin{bmatrix} 60.5 \\ 100 \end{bmatrix} (y - \hat{y}).$$

Example 6.7. Consider the transfer function of a controlled system:

$$W_0(s) = \frac{1}{s(s + 6)}.$$

Find a vector k such that the closed loop system has eigenvalues $\lambda_1^* = -1$ and $\lambda_2^* = -1.5$ by state feedback, and design a full dimensional state estimator that can realize the above feedback.

Solution. Step 1. From the transfer function above, we know that the system is controllable and observable. Therefore, the state feedback matrix and estimator can be designed independently due to the separation principle.

Step 2: Design the state feedback matrix K.

For a convenient estimator design, we use the observable canonical form Π of the system directly:

$$\dot{x} = \begin{pmatrix} 0 & 0 \\ 1 & -6 \end{pmatrix} x + \begin{pmatrix} 1 \\ 0 \end{pmatrix} u$$

$$y = \begin{pmatrix} 0 & 1 \end{pmatrix} x.$$

Step 3. We have:

$$f(\lambda) = |\lambda I - A| = \lambda^2 + 6\lambda,$$

and:

$$f^*(\lambda) = (\lambda - \lambda_1^*)(\lambda - \lambda_2^*) = \lambda^2 + 8\lambda + 52$$

Here we use equation (6.37) and (6.23):

$$T_{cl} = \begin{bmatrix} Ab & b \end{bmatrix} \begin{bmatrix} 1 & 0 \\ 6 & 1 \end{bmatrix} = \begin{bmatrix} 0 & 1 \\ 1 & 0 \end{bmatrix} \begin{bmatrix} 1 & 0 \\ 6 & 1 \end{bmatrix} = \begin{bmatrix} 6 & 1 \\ 1 & 0 \end{bmatrix},$$

and:

$$\overline{K} = \begin{bmatrix} a_0 - a_0^* & a_1 - a_1^* \end{bmatrix} = \begin{bmatrix} 0 - 52 & 6 - 8 \end{bmatrix} = \begin{bmatrix} -52 & -2 \end{bmatrix}.$$

Therefore:

$$T_{cI}^{-1} = \begin{bmatrix} 6 & 1 \\ 1 & 0 \end{bmatrix}^{-1} = \begin{bmatrix} 0 & 1 \\ 1 & -6 \end{bmatrix} .$$

Hence:

$$k = -\overline{K}T_{cI}^{-1} = -\begin{bmatrix} -52 & -2 \end{bmatrix} \begin{bmatrix} 0 & 1 \\ 1 & -6 \end{bmatrix} = \begin{bmatrix} 2 & 40 \end{bmatrix} .$$

Step 4: Design the full dimensional estimator.
Suppose $G = [g_1 \; g_2]^T$, then:

$$A - GC = \begin{pmatrix} 0 & 0 \\ 1 & -6 \end{pmatrix} - \begin{pmatrix} g_1 \\ g_2 \end{pmatrix} (0 \; 1) = \begin{pmatrix} 0 & -g_1 \\ 1 & -6 - g_2 \end{pmatrix} ,$$

and:

$$\det[\lambda I - (A - GC)] = \det \begin{pmatrix} \lambda & g_1 \\ -1 & \lambda + 6 + g_2 \end{pmatrix} = \lambda^2 + (6 + g_2)\lambda + g_1 .$$

Comparing with:

$$f^*(\lambda) = (\lambda - \lambda_1^*)(\lambda - \lambda_2^*) = \lambda^2 + 20\lambda + 100$$

we can obtain:

$$G = \begin{pmatrix} 100 \\ 4 \end{pmatrix} .$$

The full dimensional estimator equation is:

$$\dot{\hat{x}} = (A - GC)\hat{x} + Gy + bu$$

$$= \begin{pmatrix} 0 & -100 \\ 1 & -20 \end{pmatrix} \hat{x} + \begin{pmatrix} 100 \\ 4 \end{pmatrix} y + \begin{pmatrix} 1 \\ 0 \end{pmatrix} u .$$

Example 6.8. The state space model of a system is:

$$\begin{cases} \dot{X} = \begin{bmatrix} 0 & 1 \\ 3 & 4 \end{bmatrix} X + \begin{bmatrix} 2 \\ 4 \end{bmatrix} u \\ y = \begin{bmatrix} 0 & 1 \end{bmatrix} X . \end{cases}$$

Try to construct a two-dimensional state estimator with poles to be -4 and -6. Find out the model of the stator estimator and plot the diagram of the system with state estimator.

Solution. The desired characteristic equation is:

$$(s + 4)(s + 6) = s^2 + 10s + 24 = 0, \quad |N| = \begin{vmatrix} C \\ CA \end{vmatrix} = \begin{vmatrix} 0 & 1 \\ 3 & 4 \end{vmatrix} = -3 \neq 0, \quad \text{rank } N = 2 .$$

The system is observable and the state estimator can be constructed as:

$$|sI - A + GC| = s^2 + (g_2 - 4)s + 3g_1 - 3 = 0, \quad g_1 = 9, \quad g_2 = 14 .$$

The state estimator is:

$$\dot{X}_g = (A - GC)X_g + Bu + Gy = \begin{bmatrix} 0 & -8 \\ 3 & -10 \end{bmatrix} X_g + \begin{bmatrix} 2 \\ 4 \end{bmatrix} u + \begin{bmatrix} 9 \\ 14 \end{bmatrix} y.$$

The diagram of the system is shown in Figure 6.9.

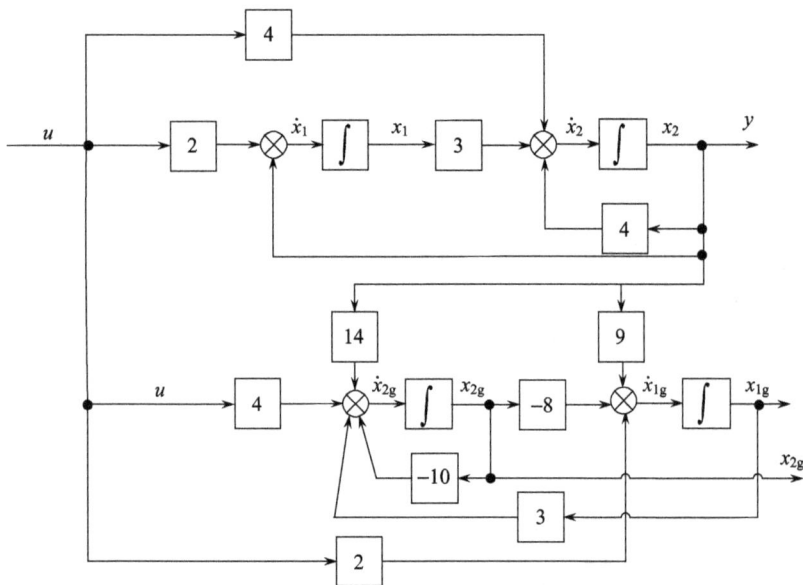

Fig. 6.9: Block diagram of the system.

Example 6.9. Consider a plant described by the following state equation:

$$\dot{x} = \begin{bmatrix} 0 & 1 & 0 \\ 0 & 0 & 1 \\ 1.24 & 0.3965 & -3.145 \end{bmatrix} x + \begin{bmatrix} 0 \\ 0 \\ 1.244 \end{bmatrix} u$$

$$y = \begin{bmatrix} 1 & 0 & 0 \end{bmatrix} x.$$

Try to design a state estimator to place the eigenvalues at $-5 \pm j5\sqrt{3}, -10$ by MATLAB.

For the example, we type:

```
a=[0 1 0;0 0 1;1.244 0.3965 -3.145];
b=[0;0;1.244];
c=[1 0 0];
v=[-5+j*5*sqrt(3) -5-j*5*sqrt(3) -10];
l=(acker(a',c',v))'
```

yield

$$l =$$
  16.8550
  147.3875
  544.3932

Then we can get:

$$L = \begin{bmatrix} 16.855 \\ 147.3875 \\ 544.3932 \end{bmatrix}.$$

So the state estimator is:

$$\dot{x} = (\Lambda - LC)x + Bu + Ly$$

$$= \left\{ \begin{bmatrix} 0 & 1 & 0 \\ 0 & 0 & 1 \\ 1.244 & 0.3965 & -3.145 \end{bmatrix} - \begin{bmatrix} 16.855 \\ 147.3875 \\ 544.3932 \end{bmatrix} \begin{bmatrix} 1 & 0 & 0 \end{bmatrix} \right\} x + \begin{bmatrix} 0 \\ 0 \\ 1.244 \end{bmatrix} u$$

$$+ \begin{bmatrix} 16.855 \\ 147.3875 \\ 544.3932 \end{bmatrix} y$$

$$= \begin{bmatrix} -16.855 & 1 & 0 \\ -147.3875 & 0 & 1 \\ -544.3932 & 0.3965 & -3.145 \end{bmatrix} x + \begin{bmatrix} 0 \\ 0 \\ 1.244 \end{bmatrix} u + \begin{bmatrix} 16.855 \\ 147.3875 \\ 544.3932 \end{bmatrix} y.$$

2. Reduced Dimensional State Estimator

The estimator presented above, usually called a full dimensional estimator, has the same dimension with the controlled system. Actually, the output vector y is always measurable. We can derive a part of state variables directly from y, thus reducing the dimension of the estimator.

Consider an observable system, assume the rank of the output matrix C is m, then m dimension state variables can be acquired by the output y. The other $n - m$ dimension state variables can be acquired by a $(n - m)$ dimensional state estimator. This estimator with the output equation can then be used to estimate all n state variables. It also has a lesser dimension than the system (6.49) and is called a reduced dimensional estimator.

The controllable system is:

$$\dot{x} = Ax + Bu$$
$$y = Cx.$$

(6.54)

With rank $C = m$, the pair (A, C) is observable. The design consists of two steps.
(1) Decompose the state to $\bar{x}_1$ and $\bar{x}_2$. m dimension $\bar{x}_2$ can be derived from y while $n - m$ dimension $\bar{x}_1$ are to be observed.
(2) Construct the $(n - m)$ dimensional state estimator.

Suppose $x = T\bar{x}$:

$$\bar{A} = T^{-1}AT = \left[\begin{array}{c:c} \bar{A}_{11} & \bar{A}_{12} \\ \hdashline \bar{A}_{21} & \bar{A}_{22} \end{array}\right] \begin{array}{c} n-m \\ m \end{array},$$

$$\bar{B} = T^{-1}B = \left[\begin{array}{c} \bar{B}_1 \\ \hdashline \bar{B}_2 \end{array}\right] \begin{array}{c} n-m \\ m \end{array},$$

$$\bar{C} = CT = \begin{array}{c} [\ 0 \ \vdots \ I\ ] \\ \hline n-m \ \vdots \ m \end{array} \quad m ,$$

$$\bar{A} = T^{-1}AT = \begin{bmatrix} \bar{A}_{11} & \bar{A}_{12} \\ \bar{A}_{21} & \bar{A}_{22} \end{bmatrix}, \quad \bar{B} = \begin{bmatrix} \bar{B}_1 \\ \bar{B}_2 \end{bmatrix}, \quad \bar{C} = CT = \begin{bmatrix} 0 & I \end{bmatrix} .$$

The transform matrix T is:

$$T^{-1} = \left[\begin{array}{c} C_0 \\ \hline C \end{array}\right] \begin{array}{c} n-m \\ m \end{array}, \quad T = \begin{bmatrix} C_0 \\ C \end{bmatrix}^{-1} ,$$

where C_0 is a $(n - m) \times n$ matrix to guarantee that T is nonsingular.

$$CT = C \begin{bmatrix} C_0 \\ C \end{bmatrix}^{-1} = \begin{bmatrix} 0 & I \end{bmatrix} .$$

The state space equation can be written as:

$$\begin{bmatrix} \dot{\bar{x}}_1 \\ \dot{\bar{x}}_2 \end{bmatrix} = \begin{bmatrix} \bar{A}_{11} & \bar{A}_{12} \\ \bar{A}_{21} & \bar{A}_{22} \end{bmatrix} \begin{bmatrix} \bar{x}_1 \\ \bar{x}_2 \end{bmatrix} + \begin{bmatrix} \bar{B}_1 \\ \bar{B}_2 \end{bmatrix} u$$

$$\bar{y} = \begin{bmatrix} 0 & I \end{bmatrix} \begin{bmatrix} \bar{x}_1 \\ \bar{x}_2 \end{bmatrix} = \bar{x}_2 .$$

(6.55)

As the system (6.54) is observable, (6.55) is also observable.

From (6.55), we can see $\bar{x}_2$ can be directly detected from $\bar{y}$, and $\bar{x}_1$ can be obtained from the estimator. The decomposed system structure is shown in Figure 6.10.

The subsystem $\bar{\Sigma}_1 = (\bar{A}_{11}, \bar{A}_{12}, \bar{B}_1, 0)$ is to be reconfigured. Following the strategy of full dimension state estimator, we can duplicate $\bar{\Sigma}_1$ from (6.55) as:

$$\dot{\bar{x}}_1 = \bar{A}_{11}\bar{x}_1 + \bar{A}_{12}\bar{x}_2 + \bar{B}_1 u = \bar{A}_{11}\bar{x}_1 + M , \tag{6.56}$$

where

$$M = \bar{A}_{12}\bar{x}_2 + \bar{B}_1 u . \tag{6.57}$$

Let $Z = \bar{A}_{21}\bar{x}_2$, then $Z = \dot{\bar{x}}_2 - \bar{A}_{22}\bar{x}_2 - \bar{B}_2 u$.

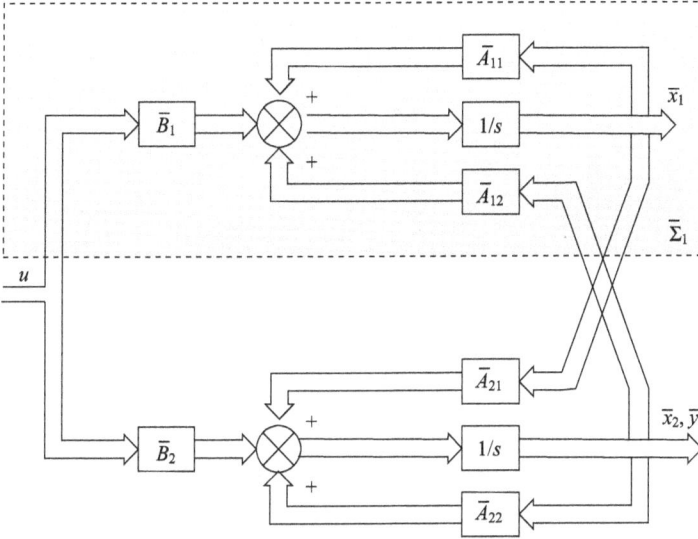

Fig. 6.10: Structure of decomposed system.

Consider M and Z as the input and output of $\overline{\Sigma}_1$, and $\overline{A}_{21}$ as the output matrix. Because of the observable pair $(\overline{A}, \overline{C})$, the pair $(\overline{A}_{11}, \overline{A}_{21})$ for $\overline{\Sigma}_1$ is also observable, thus the subsystem $\overline{\Sigma}_1$ can be estimated. Consulting equation (6.52), the estimator can be written as:

$$\dot{\widehat{\overline{x}}}_1 = (\overline{A}_{11} - \overline{G}\overline{A}_{21})\widehat{\overline{x}}_1 + M + \overline{G}Z . \tag{6.58}$$

Similarly, the eigenvalues of $(\overline{A}_{11} - \overline{G}\overline{A}_{21})$ can be assigned at desired positions by choosing a $(n - m) \times m$ dimensional matrix $\overline{G}$.

Substituting (6.57) into (6.58) yields:

$$\dot{\widehat{\overline{x}}}_1 = (\overline{A}_{11} - \overline{G}\overline{A}_{21})\widehat{\overline{x}}_1 + (\overline{A}_{12} - \overline{G}\overline{A}_{22})\overline{y} + (\overline{B}_1 - \overline{G}\overline{B}_2)u + \overline{G}\dot{\overline{y}} .$$

Considering the difficulty of implementation of $\dot{\overline{y}}$, we introduce a new variable of:

$$\widehat{\overline{w}} = \widehat{\overline{x}}_1 - \overline{G}\overline{y} .$$

So the estimator equation can be described as:

$$\dot{\widehat{\overline{w}}} = (\overline{A}_{11} - \overline{G}\overline{A}_{21})\widehat{\overline{x}}_1 + (\overline{A}_{12} - \overline{G}\overline{A}_{22})\overline{y} + (\overline{B}_1 - \overline{G}\overline{B}_2)u ,$$
$$\widehat{\overline{x}}_1 = \widehat{\overline{w}} + \overline{G}\overline{y} . \tag{6.59}$$

Hence, all n state variables $\widehat{\overline{x}}$ can be constructed as:

$$\widehat{\overline{x}} = \begin{bmatrix} \widehat{\overline{x}}_1 \\ \widehat{\overline{x}}_2 \end{bmatrix} = \begin{bmatrix} \widehat{\overline{w}} + \overline{G}\overline{y} \\ \overline{y} \end{bmatrix} = \begin{bmatrix} I \\ 0 \end{bmatrix} \widehat{\overline{w}} + \begin{bmatrix} \overline{G} \\ I \end{bmatrix} \overline{y} .$$

Then transform $\widehat{\overline{x}}$ to $\widehat{x}$, and we have $\overline{x} = T\widehat{x}$.

The whole structure of the estimator is shown in Figure 6.11.

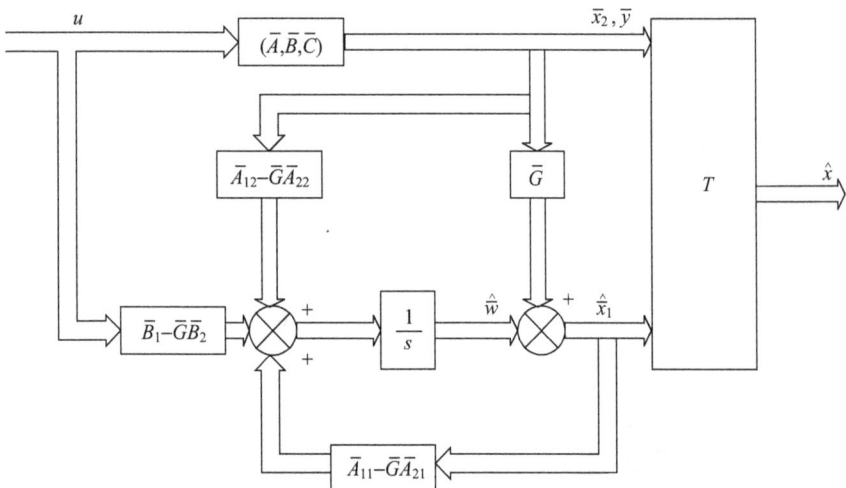

Fig. 6.11: Block diagram of reduced dimensional estimator.

From equation (6.57), we can see that $\bar{x}_2 = \bar{y}$, so there are no estimated errors of these m dimension state variables. Subtracting (6.59) from (6.56), we can obtain the estimated error equation:

$$\dot{e}_1 = \dot{\bar{x}}_1 - \dot{\hat{\bar{x}}}_1 = \bar{A}_{11}\bar{x}_1 + \bar{A}_{12}y + \bar{B}_1 u - (\bar{A}_{11} - \bar{G}\bar{A}_{21})\bar{x}_1 - (\bar{A}_{12} - \bar{G}\bar{A}_{22})\bar{y} - (\bar{B}_1 - \bar{G}\bar{B}_2)u - \bar{G}\dot{y}.$$

Considering $\bar{A}_{21}\bar{x}_1 = \dot{\bar{y}} - \bar{A}_{22}\bar{y} - \bar{B}_2 u$, the above equation can be simplified as:

$$\dot{e}_1 = (\bar{A}_{11} - \bar{G}\bar{A}_{21})(\bar{x}_1 - \hat{\bar{x}}_1) = (\bar{A}_{11} - \bar{G}\bar{A}_{21})e_1 , \tag{6.60}$$

where e_1 is the error between $\bar{x}$ and $\hat{\bar{x}}_1$. As the subsystem $\bar{\Sigma}_1$ is observable, the eigenvalues of $(\bar{A}_{11} - \bar{G}\bar{A}_{21})$ can be assigned at desired positions by choosing $\bar{G}$, thus guaranteeing that the error e_1 can approach zero at the desired rate.

Example 6.10. Consider the system:

$$\dot{x} = \begin{bmatrix} 4 & 4 & 4 \\ -11 & -12 & -12 \\ 13 & 14 & 13 \end{bmatrix} x + \begin{bmatrix} 1 \\ -1 \\ 0 \end{bmatrix} u$$

$$y = \begin{bmatrix} 1 & 1 & 1 \end{bmatrix} x .$$

Find a reduced dimensional state estimator with the poles $-3, -4$.

Solution. Examine the system's observability. There exists a reduced dimensional state estimator rank $c = 1$.

Construct the transform matrix T:

$$T^{-1} = \begin{bmatrix} 1 & 0 & 0 \\ 0 & 1 & 0 \\ 1 & 1 & 1 \end{bmatrix}, \quad T = \begin{bmatrix} 1 & 0 & 0 \\ 0 & 1 & 0 \\ -1 & -1 & 1 \end{bmatrix}.$$

Suppose:

$$\bar{A} = T^{-1}AT = \begin{bmatrix} 1 & 0 & 0 \\ 0 & 1 & 0 \\ 1 & 1 & 1 \end{bmatrix} \begin{bmatrix} 4 & 4 & 4 \\ -11 & -12 & -12 \\ 13 & 14 & 13 \end{bmatrix} \begin{bmatrix} 1 & 0 & 0 \\ 0 & 1 & 0 \\ -1 & -1 & 1 \end{bmatrix} = \begin{bmatrix} 0 & 0 & 4 \\ 1 & 0 & -12 \\ 1 & 1 & 5 \end{bmatrix},$$

we have:

$$\bar{b} = T^{-1}b = \begin{bmatrix} 1 & 0 & 0 \\ 0 & 1 & 0 \\ 1 & 1 & 1 \end{bmatrix} \begin{bmatrix} 1 \\ -1 \\ 0 \end{bmatrix} = \begin{bmatrix} 1 \\ -1 \\ 0 \end{bmatrix},$$

$$\bar{c} = cT = \begin{bmatrix} 1 & 1 & 1 \end{bmatrix} \begin{bmatrix} 1 & 0 & 0 \\ 0 & 1 & 0 \\ -1 & -1 & 1 \end{bmatrix} = \begin{bmatrix} 0 & 0 & 1 \end{bmatrix}.$$

Since $\bar{x}_3$ can be provided directly by $\bar{y}$, a second dimensional state estimator is needed.

Step 1. Introducing feedback matrix $\bar{G} = [\bar{g}_1 \; \bar{g}_2]^T$; the characteristic polynomial of the estimator yields:

$$f(\lambda) = |\lambda I - (\bar{A}_{11} - \bar{G}\bar{A}_{21})| = \left| \begin{bmatrix} \lambda & 0 \\ 0 & \lambda \end{bmatrix} - \begin{bmatrix} 0 & 0 \\ 1 & 0 \end{bmatrix} - \begin{bmatrix} \bar{g}_1 \\ \bar{g}_2 \end{bmatrix} \begin{bmatrix} 1 & 1 \end{bmatrix} \right|$$

$$= \begin{vmatrix} \lambda + \bar{g}_1 & \bar{g}_1 \\ -1 + \bar{g}_2 & \lambda + \bar{g}_2 \end{vmatrix} = \lambda^2 + (\bar{g}_1 + \bar{g}_2)\lambda + \bar{g}_1 .$$

Step 2. The desired characteristic polynomial is:

$$f^*(\lambda) = (\lambda + 3)(\lambda + 4) = \lambda^2 + 7\lambda + 12 .$$

Step 3. Comparing the corresponding coefficient of $f(\lambda)$ and $f^*(\lambda)$, we have:

$$\bar{g}_1 = 12, \bar{g}_2 = -5, \quad \text{and} \quad \bar{G} = \begin{bmatrix} 12 \\ -5 \end{bmatrix}.$$

Step 4. From equation (6.59), we obtain the estimator equation:

$$\dot{\hat{w}} = \begin{bmatrix} -12 & -12 \\ 6 & 5 \end{bmatrix} \hat{\bar{x}}_1 + \begin{bmatrix} -56 \\ 13 \end{bmatrix} \bar{y} + \begin{bmatrix} 1 \\ -1 \end{bmatrix} u$$

$$\hat{\bar{x}}_1 = \hat{w} + \begin{bmatrix} 12 \\ -5 \end{bmatrix} \bar{y} .$$

The estimation of the state after linear transformation is:

$$\hat{\overline{x}} = \begin{bmatrix} \hat{\overline{x}}_1 \\ \hat{\overline{x}}_3 \end{bmatrix} = \begin{bmatrix} \hat{\overline{w}} + \overline{G}\overline{y} \\ \overline{y} \end{bmatrix} = \begin{bmatrix} 1 & 0 \\ 0 & 1 \\ 0 & 0 \end{bmatrix} \begin{bmatrix} \overline{w}_1 \\ \overline{w}_2 \end{bmatrix} + \begin{bmatrix} 12 \\ -5 \\ 1 \end{bmatrix} \overline{y} = \begin{bmatrix} \overline{w}_1 + 12\overline{y} \\ \overline{w}_2 - 5\overline{y} \\ \overline{y} \end{bmatrix}.$$

Step 5. To get the state estimation of the original system, transform $\hat{\overline{x}}$ as follows:

$$\hat{x} = T\hat{\overline{x}} = \begin{bmatrix} 1 & 0 & 0 \\ 0 & 1 & 0 \\ -1 & -1 & 1 \end{bmatrix} \begin{bmatrix} \overline{w}_1 + 12\overline{y} \\ \overline{w}_2 - 5\overline{y} \\ \overline{y} \end{bmatrix} = \begin{bmatrix} \overline{w}_1 + 12\overline{y} \\ \overline{w}_2 - 5\overline{y} \\ -\overline{w}_1 - \overline{w}_2 - 6\overline{y} \end{bmatrix}.$$

Example 6.11. The state space model of a system is

$$\begin{cases} \dot{X} = \begin{bmatrix} 0 & 2 \\ 1 & 3 \end{bmatrix} X + \begin{bmatrix} 1 \\ 3 \end{bmatrix} u \\ y = \begin{bmatrix} 0 & 1 \end{bmatrix} X. \end{cases}$$

Try to construct a one dimensional state estimator with pole to be -5, and plot the diagram of the system.

Solution. The system model is an observable canonical, so the system is observable, the state estimator can be constructed and the pole can be assigned arbitrary $y = x_2$. Only x_1 needs to be constructed:

$$\left| sI - \hat{A}_{11} + g\hat{A}_{21} \right| = s - 0 + g = s + 5 = 0, \quad g = 5;$$

$$\dot{W} = \dot{w} = (A_{11} - GA_{21})W + [(A_{12} - GA_{22}) + (A_{11} - GA_{21})G]Y + (B_1 - GB_2)U$$
$$= -5w - 38y - 14u,$$

$$x_{1g} = \hat{x}_{1g} = W + GY = w + 5y, \quad x_{2g} = \hat{x}_{2g} = y.$$

The state estimator is:

$$\begin{cases} \dot{w} = -5w - 38y - 14u \\ x_{1g} = w + 5y \\ x_{2g} = y. \end{cases}$$

The diagram of the system with state estimator is shown in Figure 6.12.

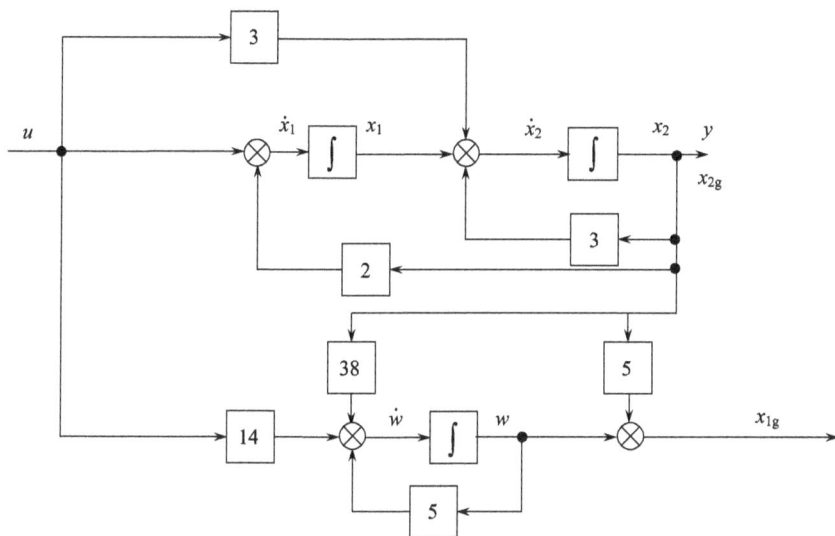

Fig. 6.12: Diagram of the system.

6.5 State Feedback Based on State Estimator

Figure 6.13 is a state feedback system based on full dimensional state estimator.
Consider the controllable and observable controlled system $\Sigma_0 = (A, B, C)$:

$$\left.\begin{array}{l} \dot{x} = Ax + Bu \\ y = Cx \end{array}\right\} \qquad (6.61)$$

The state estimator Σ_G:

$$\left.\begin{array}{l} \dot{\hat{x}} = (A - GC)\hat{x} + Gy + Bu \\ \hat{y} = C\hat{x} \end{array}\right\} \qquad (6.62)$$

The state feedback law is:

$$u = -K\hat{x} + v . \qquad (6.63)$$

By substituting equation (6.63) into equation (6.61) and equation (6.62), you can obtain
the state space description of the total closed loop system.

$$\left.\begin{array}{l} \dot{x} = Ax - BK\hat{x} + Bv \\ \dot{\hat{x}} = GCx + (A - GC - BK)\hat{x} + Bv \\ y = Cx \end{array}\right\} \qquad (6.64)$$

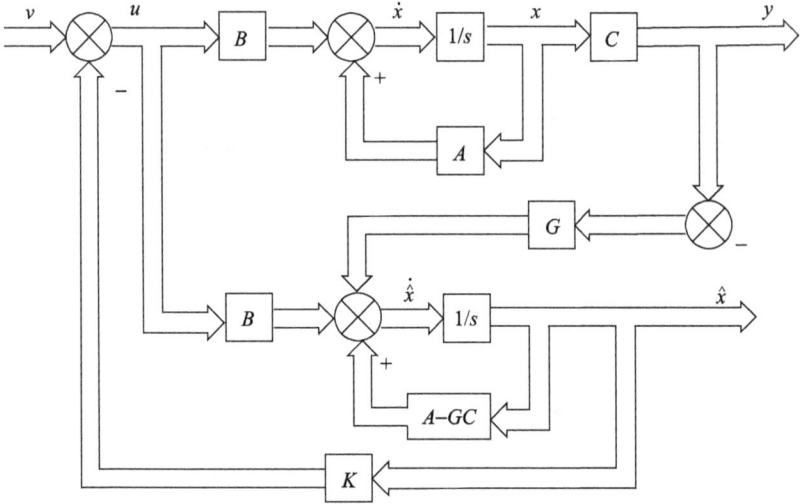

Fig. 6.13: State feedback system based on full dimensional state estimator.

Equation (6.64) can be written in the following matrix form:

$$\left. \begin{array}{l} \begin{pmatrix} \dot{x} \\ \dot{\hat{x}} \end{pmatrix} = \begin{pmatrix} A & -BK \\ GC & A - GC - BK \end{pmatrix} \begin{pmatrix} x \\ \hat{x} \end{pmatrix} + \begin{pmatrix} B \\ B \end{pmatrix} v \\ \\ y = \begin{pmatrix} C & 0 \end{pmatrix} \begin{pmatrix} x \\ \hat{x} \end{pmatrix} \end{array} \right\}$$

(6.65)

This is a closed loop system with dimension of $2n$.

Define the state error as $\tilde{x} = x - \hat{x}$ and introduce the following equivalent transformation:

$$\begin{pmatrix} x \\ \tilde{x} \end{pmatrix} = \begin{pmatrix} I & 0 \\ I & -I \end{pmatrix} \begin{pmatrix} x \\ \hat{x} \end{pmatrix} = \begin{pmatrix} x \\ x - \hat{x} \end{pmatrix} .$$

(6.66)

Suppose the transfer matrix is:

$$T = \begin{pmatrix} I & 0 \\ I & -I \end{pmatrix} .$$

(6.67)

Then:

$$T^{-1} = \begin{pmatrix} I & 0 \\ I & -I \end{pmatrix}^{-1} = \begin{pmatrix} I & 0 \\ I & -I \end{pmatrix} = T .$$

(6.68)

With linear transformation, the system turns into $(\overline{A}_1, \overline{B}_1, \overline{C}_1)$:

$$\overline{A}_1 = T^{-1}A_1 T = \begin{pmatrix} I & 0 \\ I & -I \end{pmatrix}\begin{pmatrix} A & -BK \\ GC & A - GC - BK \end{pmatrix}\begin{pmatrix} I & 0 \\ I & -I \end{pmatrix} = \begin{pmatrix} A - BK & BK \\ 0 & A - GC \end{pmatrix}$$

$$\overline{B}_1 = T^{-1}B_1 = \begin{pmatrix} I & 0 \\ I & -I \end{pmatrix}\begin{pmatrix} B \\ B \end{pmatrix} = \begin{pmatrix} B \\ 0 \end{pmatrix} \tag{6.69}$$

$$\overline{C}_1 = C_1 T = \begin{pmatrix} C & 0 \end{pmatrix}\begin{pmatrix} I & 0 \\ I & -I \end{pmatrix} = \begin{pmatrix} C & 0 \end{pmatrix}.$$

Linear transformation does not change the poles of system, therefore:

$$\det\left[\lambda I - \overline{A}_1\right] = \det\begin{pmatrix} \lambda I - (A - BK) & -BK \\ 0 & \lambda I - (A - GC) \end{pmatrix}$$

$$= \det[\lambda I - (A - BK)]\det[\lambda I - (A - GC)] \tag{6.70}$$

$$= \det[\lambda I - (A - BK)]\det[\lambda I - (A - GC)].$$

The results are very interesting, because they illustrate the fact that the characteristic polynomial of the closed loop state feedback system based on state estimator equals the product of the characteristic polynomial of the matrix $(A - BK)$, and that of the matrix $(A - GC)$. The poles of the closed loop system equals the sum of the poles of direct state feedback $(A - BK)$ and that of state estimator $(A - GC)$. Indeed, if system (A, B) is controllable, then the matrix K of the state feedback law (6.63) can be chosen so that the poles of the closed loop system $\Sigma_0 = (A, B, C)$ have any desired arbitrary values. The same applies to equation (6.62) where, if the system (A, C) is observable, the matrix G of estimator can be chosen so as to force the error to go rapidly to zero. This property, where the two design problems (the estimator and the matrix K of the closed loop system) can be handled independently, is called the *separation principle*. This principle is clearly a very important design feature, since it reduces a very difficult design task to two separate simpler design problem.

Consider the pole assignment and the estimator design problem. The pole assignment problem is called the *control problem* and it is rather a simple control design tool for improving the closed loop system performance. The estimator design problem is called *estimator problem*, since it produces a good estimate of $x(t)$ in cases where $x(t)$ is not measurable. The solution of the estimator design problem reduces to that of solving a pole assignment problem. In cases where an estimate of $x(t)$ is used in the control problem, one faces the problem of simultaneously solving the estimation and the control problem. At first sight this appears to be a formidable task. However, thanks to the separation theorem, the solution of the combined problem of estimation and control breaks down to separately solving the estimation and the control problem. The solution of the combined problem of estimation and control requires twice of the pole assignment.

6.6 Summary

Three types of feedback are introduced in this chapter. They can be used to improve the performance of a system. The precondition and algorithm of every feedback are discussed in detail. The pole assignment can be realized with some feedback. The desired poles come from the request for the performance of a system. The state estimator can be designed when a system is observable to realize the state estimation so as to fulfill the state feedback to optimize the system performance.

Appendix: State Feedback and Observer for Main Steam Temperature Control in Power Plant Steam Boiler Generation System

The superheater is an important part of the steam generation process in the boiler turbine system, where steam is superheated before entering the turbine that drives the generator. The objective is to control the superheated steam temperature by controlling the flow of spray water using the spray water valves. As can be seen in Figure 6.14, a two stage water sprayer is used to control the superheated temperature. The steam generated from the boiler drum passes through the low temperature superheater before it enters the radiant type platen superheater. Water is sprayed onto the steam to control the superheated steam temperature in both the low and high temperature superheaters. Proper control of the superheated steam temperature is extremely important to ensure the overall efficiency and safety of the power plant. Therefore, the superheated steam temperature is to be controlled by adjusting the flow of spray water.

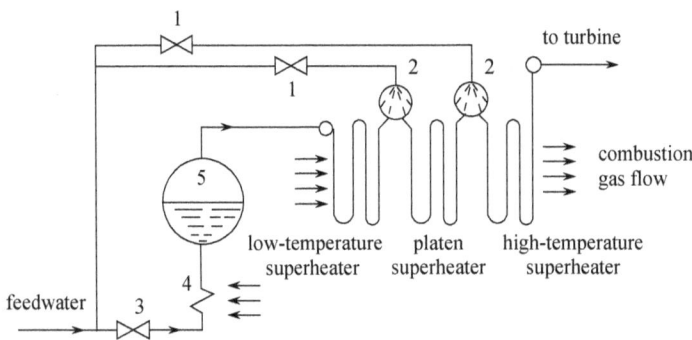

Fig. 6.14: Boiler and superheater steam generation process.

The typical mathematic model of the superheated steam temperature control process is a sixth order transfer function as follows:

$$G_0(s) = \frac{\theta(s)}{W(s)} = \frac{1.589 \times 2.45}{(1 + 14s)^2(1 + 15.8s)^4},$$

where θ and W represent the superheater steam temperature and the water flow rate of spray superheating, respectively.

Then the transfer function can be transformed to the controllable canonical form:

$$\dot{x}(t) = Ax(t) + Bu(t)$$
$$y(t) = Cx(t) + Du(t).$$

where:

$$A = \begin{bmatrix} -0.396 & -0.0653 & -0.0057 & -2.8355e-6 & -7.4664e-6 & -8.1868e-8 \\ 1 & 0 & 0 & 0 & 0 & 0 \\ 0 & 1 & 0 & 0 & 0 & 0 \\ 0 & 0 & 1 & 0 & 0 & 0 \\ 0 & 0 & 0 & 1 & 0 & 0 \\ 0 & 0 & 0 & 0 & 1 & 0 \end{bmatrix},$$

$$B = \begin{bmatrix} 1 \\ 0 \\ 0 \\ 0 \\ 0 \\ 0 \end{bmatrix}, \quad C = \begin{bmatrix} 0 & 0 & 0 & 0 & 0 & 0.3187e-6 \end{bmatrix}.$$

Here, we need to find a state feedback controller $u = r - [k_1 \ k_2 \ k_3 \ k_4 \ k_5 \ k_6]x$ to make the closed loop poles at $[-0.1 \ -0.1 \ -0.1 \ -0.1 \ -0.1 \ -0.1]$.

The block diagram of the system is shown in Figure 6.15.

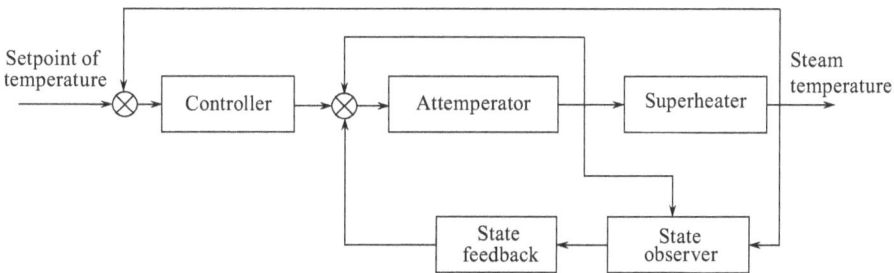

Fig. 6.15: The steam temperature control system with state feedback and state observer.

The system is controllable according to the matrix A. With state feedback, the closed loop characteristic polynomial is:

$$f(\lambda) = \det[\lambda I - (A - BK)]$$
$$= \lambda^6 + (0.396 + k_1)\lambda^5 + (0.0653 + k_2)\lambda^4 + (0.0057 + k_3)\lambda^3$$
$$+ (2.8355e{-}6 + k_4)\lambda^2 + (7.4664e{-}6 + k_5)\lambda + 8.1868e{-}8 + k_6 .$$

The desired closed loop characteristic polynomial is:

$$f^*(\lambda) = (\lambda + 0.1)^6 = \lambda^6 + 0.6\lambda^5 + 0.15\lambda^4 + 0.02\lambda^3 + 0.0015\lambda^2 + 0.00006\lambda + 0.000001 .$$

Compare relative parameters in the above two functions, and we have:

$$k_1 = 0.204 , \qquad k_2 = 0.0847 , \qquad k_3 = 0.0143 ,$$
$$k_4 = 1.4972e{-}3 , \qquad k_5 = 5.2534e{-}5 , \qquad k_6 = 9.1813e{-}7 .$$

Thus:

$$K = \begin{bmatrix} 0.204 & 0.0847 & 0.0143 & 1.4972e{-}3 & 5.2534e{-}5 & 9.1813e{-}7 \end{bmatrix} .$$

The observable canonical form of the system is:

$$\dot{x}(t) = \overline{A}x(t) + \overline{B}u(t)$$
$$y(t) = \overline{C}x(t) + \overline{D}u(t) ,$$

where:

$$\overline{A} = \begin{bmatrix} 0 & 0 & 0 & 0 & 0 & -8.1868e{-}8 \\ 1 & 0 & 0 & 0 & 0 & -7.4664e{-}6 \\ 0 & 1 & 0 & 0 & 0 & -2.8355e{-}6 \\ 0 & 0 & 1 & 0 & 0 & -0.0057 \\ 0 & 0 & 0 & 1 & 0 & -0.0653 \\ 0 & 0 & 0 & 0 & 1 & -0.396 \end{bmatrix} ,$$

$$\overline{B} = \begin{bmatrix} 0.3187e{-}6 \\ 0 \\ 0 \\ 0 \\ 0 \\ 0 \end{bmatrix} , \qquad \overline{C} = \begin{bmatrix} 0 & 0 & 0 & 0 & 0 & 1 \end{bmatrix} .$$

Next, we could design a full dimensional state estimator with poles to be $[-0.25\ -0.25\ -0.25\ -0.25\ -0.25\ -0.25]$.

Design the full dimensional estimator.

Suppose $G = [g_1 \ g_2 \ g_3 \ g_4 \ g_5 \ g_6]^{\mathrm{T}}$, then:

$$\overline{A} - G\overline{C} = \begin{bmatrix} 0 & 0 & 0 & 0 & 0 & -8.1868\text{e}{-8} \\ 1 & 0 & 0 & 0 & 0 & -7.4664\text{e}{-6} \\ 0 & 1 & 0 & 0 & 0 & -2.8355\text{e}{-6} \\ 0 & 0 & 1 & 0 & 0 & -0.0057 \\ 0 & 0 & 0 & 1 & 0 & -0.0653 \\ 0 & 0 & 0 & 0 & 1 & -0.396 \end{bmatrix} - \begin{bmatrix} g_1 \\ g_2 \\ g_3 \\ g_4 \\ g_5 \\ g_6 \end{bmatrix} \begin{bmatrix} 0 & 0 & 0 & 0 & 0 & 1 \end{bmatrix}$$

$$= \begin{bmatrix} 0 & 0 & 0 & 0 & 0 & -8.1868\text{e}{-8} - g_1 \\ 1 & 0 & 0 & 0 & 0 & -7.4664\text{e}{-6} - g_2 \\ 0 & 1 & 0 & 0 & 0 & -2.8355\text{e}{-6} - g_3 \\ 0 & 0 & 1 & 0 & 0 & -0.0057 - g_4 \\ 0 & 0 & 0 & 1 & 0 & -0.0653 - g_5 \\ 0 & 0 & 0 & 0 & 1 & -0.396 - g_6 \end{bmatrix},$$

and

$$\det\left[\lambda I - \left(\overline{A} - G\overline{C}\right)\right] = \lambda^6 + (0.396 + g_6)\lambda^5 + (0.0653 + g_5)\lambda^4 + (0.0057 + g_4)\lambda^3$$
$$+ (2.8355\text{e}{-6} + g_3)\lambda^2 + (7.4664\text{e}{-6} + g_2)\lambda$$
$$+ 8.1868\text{e}{-8} + g_1 .$$

Comparing with:

$$f^*(\lambda) = (\lambda + 0.25)^6 = \lambda^6 + 1.5\lambda^5 + 0.9375\lambda^4 + 0.3125\lambda^3 + 0.0586\lambda^2$$
$$+ 0.00586\lambda + 0.000244 ,$$

we can obtain:

$$G = \begin{bmatrix} 0.0002 \\ 0.0059 \\ 0.0583 \\ 0.3068 \\ 0.8722 \\ 1.104 \end{bmatrix} .$$

The full dimensional estimator equation is:

$$\dot{\hat{x}} = \left(\overline{A} - G\overline{C}\right)\hat{x} + Gy + \overline{B}u$$

$$= \begin{bmatrix} 0 & 0 & 0 & 0 & 0 & -0.000244 \\ 1 & 0 & 0 & 0 & 0 & -0.00586 \\ 0 & 1 & 0 & 0 & 0 & -0.0586 \\ 0 & 0 & 1 & 0 & 0 & -0.3125 \\ 0 & 0 & 0 & 1 & 0 & -0.9375 \\ 0 & 0 & 0 & 0 & 1 & -1.5 \end{bmatrix} \hat{x} + \begin{bmatrix} 0.0002 \\ 0.0059 \\ 0.0583 \\ 0.3068 \\ 0.8722 \\ 1.104 \end{bmatrix} y + \begin{bmatrix} 0.3187\text{e}{-6} \\ 0 \\ 0 \\ 0 \\ 0 \\ 0 \end{bmatrix} u .$$

Exercise

6.1. Determine whether the following systems can realize arbitrary pole assignment with state feedback:

$$(1) \quad \dot{x} = \begin{bmatrix} 1 & 2 \\ 3 & 1 \end{bmatrix} x + \begin{bmatrix} 1 \\ 0 \end{bmatrix} u$$

$$(2) \quad \dot{x} = \begin{bmatrix} 4 & 2 \\ 0 & -2 \end{bmatrix} x + \begin{bmatrix} 1 \\ 0 \end{bmatrix} u$$

$$(3) \quad \dot{x} = \begin{bmatrix} 1 & 0 & 0 \\ 0 & -2 & 1 \\ 0 & 0 & -2 \end{bmatrix} x + \begin{bmatrix} 1 & 0 \\ 0 & 1 \\ 0 & 0 \end{bmatrix} u$$

$$(4) \quad \dot{x} = \begin{bmatrix} 0 & 1 & 0 & 0 \\ 0 & 0 & 1 & 0 \\ 0 & 0 & 0 & 1 \\ -2 & -4 & -3 & -5 \end{bmatrix} x + \begin{bmatrix} 0 & 0 & 0 \\ 0 & 0 & 1 \\ 0 & 1 & 0 \\ 1 & 0 & 0 \end{bmatrix} u$$

6.2. Consider a single input continuous time LTI system:

$$\dot{x} = \begin{bmatrix} 1 & 2 \\ 3 & 1 \end{bmatrix} x + \begin{bmatrix} 1 \\ 0 \end{bmatrix} u.$$

Try to find a state feedback matrix k, which makes the closed loop eigenvalues $\lambda_1^* = -2 + j$, $\lambda_2^* = -2 - j$.

6.3. Given the transfer function of a SISO continuous time LTI system:

$$G(s) = \frac{1}{s(s+4)(s+8)},$$

try to find a state feedback matrix k, which makes the closed loop eigenvalues $\lambda_1^* = -2$, $\lambda_2^* = -4$, $\lambda_3^* = -7$.

6.4. Given a single input LTI system:

$$\dot{x} = \begin{bmatrix} 0 & 0 & 0 \\ 1 & -6 & 0 \\ 0 & 1 & -12 \end{bmatrix} x + \begin{bmatrix} 1 \\ 0 \\ 0 \end{bmatrix} u,$$

Try to find a state feedback matrix $u = -Kx$, which makes the closed loop eigenvalues $\lambda_1^* = -2$, $\lambda_2^* = -1 + j$, $\lambda_3^* = -1 - j$.

6.5. Consider a continuous time LTI system:

$$\dot{x} = \begin{bmatrix} 1 & 1 \\ 0 & 1 \end{bmatrix} x + \begin{bmatrix} 0 \\ 1 \end{bmatrix} u$$

$$y = \begin{bmatrix} 2 & 0 \\ 0 & 1 \end{bmatrix} x.$$

Try to find an output feedback matrix f, which makes the closed loop eigenvalues become $\lambda_1^* = -2$, $\lambda_2^* = -4$.

6.6. Consider the following fourth order system:

$$\dot{x} = \begin{bmatrix} 2 & 1 & 0 & 0 \\ 0 & 2 & 0 & 0 \\ 0 & 0 & -2 & 0 \\ 0 & 0 & 0 & -2 \end{bmatrix} x + \begin{bmatrix} 0 \\ 1 \\ 1 \\ 1 \end{bmatrix} u .$$

Determine state feedback matrix to place the closed loop system poles at:

(1) $\lambda_1^* = -2$, $\lambda_2^* = -2$, $\lambda_3^* = -2$, $\lambda_3^* = -2$

(2) $\lambda_1^* = -3$, $\lambda_2^* = -3$, $\lambda_3^* = -3$, $\lambda_3^* = -2$

(3) $\lambda_1^* = -3$, $\lambda_2^* = -4$, $\lambda_3^* = -3$, $\lambda_3^* = -3$

6.7. Consider a continuous time LTI system:

$$\dot{x} = \begin{bmatrix} 1 & 1 & 0 \\ 0 & 1 & 0 \\ 0 & 0 & 2 \end{bmatrix} x + \begin{bmatrix} 0 & 0 \\ 1 & 0 \\ 0 & -1 \end{bmatrix} u .$$

Try to find the state feedback matrix to place the closed loop eigenvalues at $\lambda_1^* = -2$, $\lambda_2^* = -1 + j2$, $\lambda_3^* = -1 - j2$.

6.8. Given the transfer function of a SISO continuous time LTI system:

$$g_0(s) = \frac{(s + 2)(s + 3)}{(s + 1)(s - 2)(s + 4)} ,$$

try to determine if there exists a state feedback matrix k, which can make the closed loop transfer function as:

$$g(s) = \frac{s + 3}{(s + 2)(s + 4)} .$$

If it does, find a state feedback matrix k.

6.9. Design a full dimensional state estimator with eigenvalues to at $-r$, $-2r$ ($r > 0$) for the state equation below:

$$\dot{x} = \begin{bmatrix} 0 & 1 \\ 0 & 0 \end{bmatrix} x + \begin{bmatrix} 0 \\ 1 \end{bmatrix} u$$

$$y = \begin{bmatrix} 1 & 0 \end{bmatrix} x .$$

6.10. Design a reduced dimensional state estimator with eigenvalues to as -4 and -5 for the state equation below:

$$\dot{x} = \begin{bmatrix} 0 & 1 & 0 \\ 0 & 0 & 1 \\ 0 & 0 & 0 \end{bmatrix} x + \begin{bmatrix} 0 \\ 0 \\ 1 \end{bmatrix} u$$

$$y = \begin{bmatrix} 1 & 0 & 0 \end{bmatrix} x .$$

Bibliography

[1] C. T. Chen. Linear System Theory and Design. Oxford University Press. 1999. (Third Edition).
[2] J. J. D'Azzo. Linear Control System Analysis and Design. McGraw–Hill, 1995.
[3] J. J. D'Azzo. Linear Control System Analysis and Design With MATLAB. Marcel Dekker, 2003, 1995.
[4] K. Ogata. Modern Control Engineering: Fourth Edition. Prentice–Hall. Inc. 2006.
[5] B. Liu. Modern Control Theory. Second Edition. China Machine Press, 1992 (only available in Chinese).
[6] P. N. Paraskevoppulos. Modern Control Engineering. Marcel Dekker. Inc 2002.
[7] D. Z. Zheng. Linear System Theory. Second Edition. Tsinghua University Press, 2002 (only available in Chinese).

https://doi.org/10.1515/9783110574951-007

Index

https://doi.org/10.1515/9783110574951-008

www.ingramcontent.com/pod-product-compliance
Lightning Source LLC
Chambersburg PA
CBHW081522220326

41598CB00036B/6298